❦ BAFFIN ISLAND ❧

NORTHERN LIGHTS SERIES

COPUBLISHED WITH THE ARCTIC INSTITUTE OF NORTH AMERICA
ISSN 1701-0004 (PRINT) ISSN 1925-2943 (ONLINE)

This series takes up the geographical region of the North (circumpolar regions within the zone of discontinuous permafrost) and publishes works from all areas of northern scholarship, including natural sciences, social sciences, earth sciences, and the humanities.

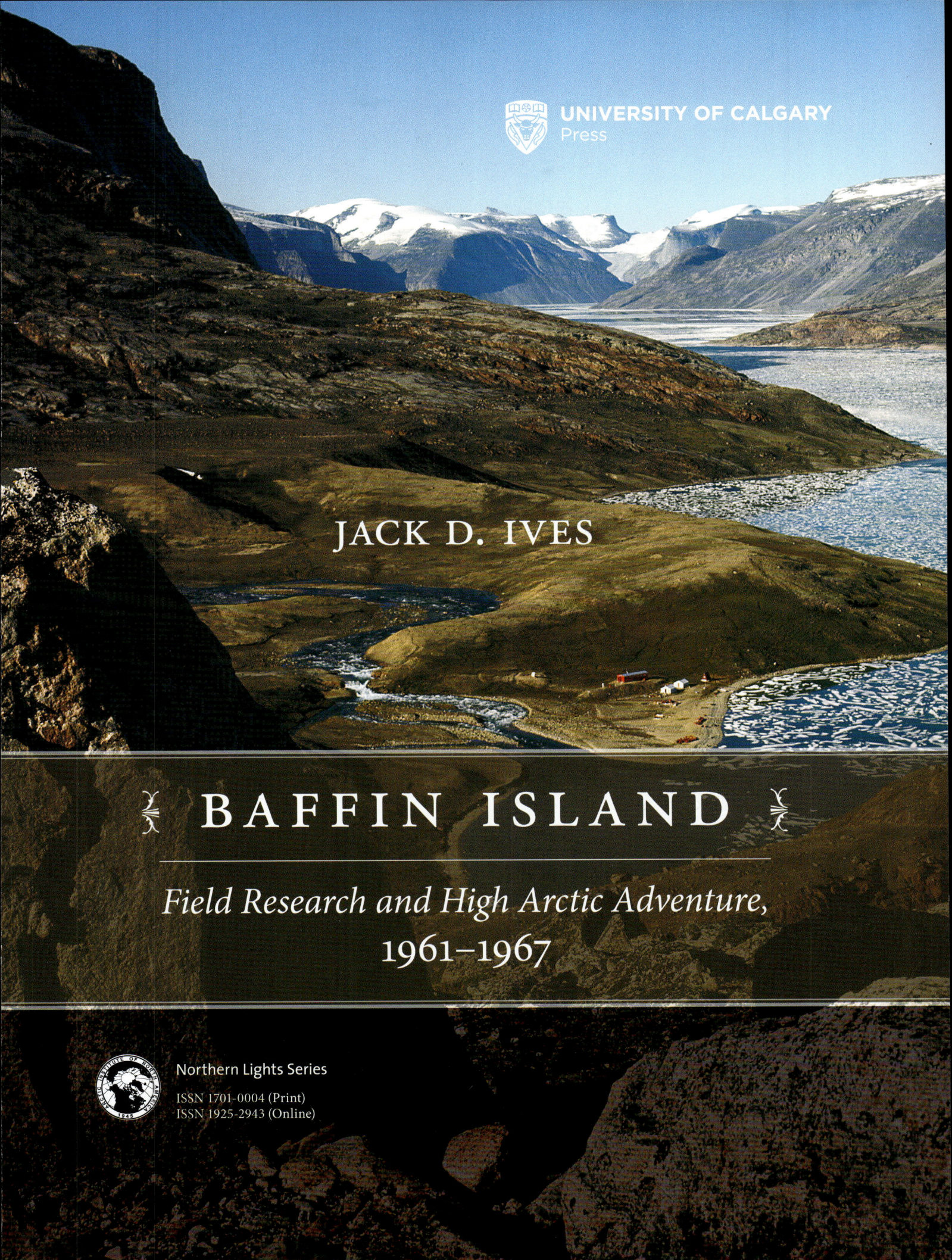

UNIVERSITY OF CALGARY
Press

JACK D. IVES

❧ BAFFIN ISLAND ❧

Field Research and High Arctic Adventure,
1961–1967

Northern Lights Series

ISSN 1701-0004 (Print)
ISSN 1925-2943 (Online)

University of Calgary Press
2500 University Drive NW
Calgary, Alberta
Canada T2N 1N4

www.uofcpress.com

LIBRARY AND ARCHIVES CANADA CATALOGUING IN PUBLICATION

Ives, Jack D., author
      Baffin Island : field research and high Arctic adventure, 1961-1967 / Jack D. Ives.

(Northern lights series, ISSN 1701-0004 ; no. 18)
Includes bibliographical references and index.
Issued in print and electronic formats.
ISBN 978-1-55238-829-7 (paperback).–ISBN 978-1-55238-831-0 (pdf).–
ISBN 978-1-55238-830-3 (open access pdf).–ISBN 978-1-55238-832-7 (epub).–
ISBN 978-1-55238-833-4 (mobi)

      1. Ives, Jack D.–Travel–Nunavut–Baffin Island. 2. Geographers–Travel–
Nunavut–Baffin Island. 3. Geographers–Canada–Biography. 4. Geography–
Fieldwork–Nunavut–Baffin Island–History–20th century. 5. Arctic regions–
Research–Canada–History–20th century. 6. Research–Government policy–
Canada–History–20th century. 7. Baffin Island (Nunavut)–History–20th
century. I. Title. II. Series: Northern lights series ; no. 18

G69.I94A3 2016   910.92   C2016-900034-6
                                     C2016-900035-4

The University of Calgary Press acknowledges the support of the Government of Alberta through the Alberta Media Fund for our publications. We acknowledge the financial support of the Government of Canada through the Canada Book Fund for our publishing activities. We acknowledge the financial support of the Canada Council for the Arts for our publishing program.

  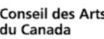

Printed and bound in Canada by Friesens
♻ This book is printed on Garda Silk paper

Copy editing and indexing by Alison Jacques
Cover design, page design, and typesetting by Melina Cusano

Front Cover
Base camp on Inugsuin Fiord at 69°40' N, 70°00' W. The view is toward the northeast along the fiord; the main building and tents (at centre) are just beyond the open water. The river, enlarged by melting snow, ensures local clearing of the fiord ice cover in the small embayment (good for swimming, with sandy beach!). Government icebreakers were able to approach within 150 metres of the beach. The camp, which was the centre of expedition operations for the 1965–1967 field seasons, lies more than one hundred kilometres inland from Baffin Bay, beyond Clyde River settlement. (Photo: July 1966)

Frontispiece
In the 1960s, the mountains, ice caps, and glaciers of the northeastern section of Baffin Island were virtually unknown to most Canadians. Even today, except to the local Inuit and a growing, but still small, number of mountaineers anxious to test their skills, the region remains a "lost" section of Canada. (Photo: July 1965)

TO

Cuchlaine and Olav

Lifelong friends
and loyal colleagues

Table of Contents

FOREWORD

BY

The Honourable Peter Adams

This book brings to life both the adventure and the rigour of Arctic research during the "golden age" of Canadian federal research, in the 1960s. Its focus is the series of field seasons when scientists from the Geographical Branch of the Department of Energy, Mines, and Resources, Canada, tackled the glacial geomorphology and glaciology of Baffin Island. As far as the North was concerned, this golden age of federal research reflected both changing attitudes and increased resources in Ottawa and changes in science and communications in the region itself. For example, completion of the air photo coverage of Canada provided a wealth of information about Canada's highest latitudes. At the same time, the improved transportation and communication provided by the Cold War–era Distant Early Warning Line (DEW Line) stations and the High Arctic Weather Stations enhanced the safety and organization of Northern Canadian field research.

The Geographical Branch seized on opportunities such as these for a large-scale, long-term study of the landscape of Baffin Island. This work particularly benefited from preliminary air photo analysis by Jack Ives and his colleagues in the Quebec–Labrador Peninsula (described in the author's memoir *The Land Beyond* [University of Alaska Press, 2010]). Great interest in deciphering the pattern of retreat of the Laurentide Ice Sheet led to a research strategy modelled on similar work describing the radial retreat of the equivalent ice sheet in Scandinavia. For Baffin Island, the model was tested through an elaborate, multi-season field exercise that involved footslogging but also innovative use of light aircraft, helicopters, and ships. However, even with a DEW Line station as anchor, air travel in the North still involved a good deal of trial and error. The field research, in various disciplines, covered much of the island.

The adventure and rigour of government-sponsored research was not confined to the work on Baffin Island. The author documents aspects of the evolution of science within the federal government in Ottawa, recounting, among other things, the rise and eventual fall of the Geographical Branch itself and the institutionalization of glaciology in the federal research system.

On Baffin Island, glaciology was well represented in the field program, through studies of the mass balance, volume, and snow stratigraphy of the Barnes Ice Cap. While earlier groups had worked on the ice caps of Baffin Island, this was the first large-scale attempt at understanding the Barnes, which was widely believed to represent a relic of the Laurentide Ice Sheet. This glacier work was undertaken at a time when, farther north, university-based expeditions began studying High Arctic glaciers on Axel Heiberg and Devon islands. All of this work was important in better understanding the nature and nourishment of high-latitude glaciers in general.

In the 1960s, the expanding universities of Canada were beginning to produce graduates trained for polar research. Ives had helped lay the groundwork for this at Schefferville's McGill Sub-Arctic Research Laboratory in the 1950s. The sixties were also a time when the concept of gender balance in field research was in its infancy. The Geographical Branch research on Baffin Island represented both a career opportunity and a training ground for male and female students, many of whom subsequently contributed greatly to polar science.

Jack Ives himself, through a lifetime of Arctic and mountain research, is uniquely qualified to weave together the threads in this book. Through early work in Scandinavia, including Iceland, and later at the University of Colorado's Institute of Arctic and Alpine Research, he links the Scandinavian, American, and Canadian views of Pleistocene science. From his graduate student and faculty days at McGill to his work as assistant director and director of the Geographical Branch, his involvement spanned the period when Canada was striving to produce home-grown Arctic researchers while also developing a federal capacity for Arctic science.

Ives's relish for Arctic fieldwork and his delight in scientific debate come alive in this gripping and informative memoir.

Peter Adams
Professor Emeritus, Trent University, Peterborough
February 2014

✠ ACKNOWLEDGMENTS ✠

Preparation of this account of a series of expeditions to Baffin Island that took place in the 1960s has depended on access to field reports, the published papers of many individuals, my own diaries, and the considerable stretch of my memory. This has led, understandably, to a heavy reliance on recollections of field colleagues, their diaries, and correspondence. Olav Løken, despite a debilitating illness, painstakingly read through an early draft and made available his thorough field reports from 1964 to 1967. Roger Barry, Michael Church, Art Dyke, John England, George Falconer, Gunnar Østrem, and Patrick Webber have read through all or parts of the text and have contributed many valuable corrections, suggestions, and additions from their own research material. Pat, in particular, supplied a large amount of his subsequent research material and its interpretation. Norma Sagar provided information about Brian, her late husband. Jane (Philpot) Buckley and Lyn (Drapier) Arsenault are acknowledged for many helpful assists. Angus Hamilton, leader of the 1961 Dominion Observatory's geophysical expedition to Baffin Island, read chapter 2 and provided a valuable check on the details of several incidents. Gifford Miller, based on his extensive and continuing Arctic research, along with John England and Art Dyke, has generously kept me informed of new developments and evolving interpretations.

Three other colleagues, not part of the Baffin Island field teams, also read and commented on an early version of the manuscript, and thus special thanks are due to Denis St-Onge, Peter McKinnon, and Pierre Camu. Dr. Camu, former president of the St. Lawrence Seaway Authority, kindly read through and extensively annotated a first draft that had set the far more complicated goal of combining the detailed administrative struggles of the Geographical Branch, of which he had been an earlier senior member, with the Baffin Island operations. He urged that the two topics should be treated separately, except for the necessary linkages and explanations. His wise advice convinced me, leaving the outline of a second book waiting in my filing cabinet.

Not only did David Harrison serve as helicopter pilot *par excellence*, field companion, advisor, and friend from 1965 to 1967, but I "rediscovered" him later in life as an accomplished writer and editor. I cannot thank him enough for reliving with me much of the challenge and excitement among Baffin Island's high mountains and fiords, although I will refer here specifically to his ability to provide very necessary technical corrections to my frequent accounts of the manner in which he was able to negotiate through oftentimes daunting flying conditions. His overall editorial perspicacity has been of immeasurable help.

The enthusiasm and general support of the entire staff of the former Geographical Branch of the Department of Mines and Technical Surveys (becoming Energy, Mines, and Resources in 1966) deserve a special mention. These individuals were the essential "home team." At the higher administrative levels, I cannot exaggerate my great respect, admiration, and gratitude for the wisdom, guidance, and encouragement afforded me by the late Dr. W. E. van Steenburgh, deputy minister. I also record my special respect and thanks for the role played by the late Dr. James Harrison, director of the Geological Survey of Canada and assistant deputy minister. Our ideological conflict, as detailed in the text, did not put a dent in his encouragement for the Baffin Island endeavour, which materialized into vital support for my later career development from his subsequent positions of assistant director-general of UNESCO and chair of the United Nations University Scientific Advisory Committee. From among the other senior departmental administrators, former Geographical Branch Director the late Dr. Norman Nicholson's initial acceptance of what must have seemed to him an overambitious personal dream provided the first essential building block.

Special thanks are due to all members of the field teams and support staff that numbered over fifty men and women, many very youthful, some of them teenagers. Their camaraderie, determination in often strenuous conditions, persistence, and staying power continue to impress me a half century later. From among them I must single out Professor Cuchlaine King, who continues to provide encouragement and inspiration, even from her extended-care home in Wensleydale, Yorkshire. She was my undergraduate tutor and the faculty member of the University of Nottingham student expeditions to Iceland. She was also the spearhead for breaking through the Canadian federal government opposition to gender equality in the Arctic.

In the present era of apparent governmental indifference to a high level of free enterprise research in the Arctic, made the more onerous by the seeming limitation of freedom of expression by governmental scientists, designation of the Baffin Island years as a golden age is perhaps appropriate. In retrospect, the Baffin Island team members of 1961 to 1967 were certainly beneficiaries of what may be classed as special privilege. Many of their results have turned out to bear acutely on the impacts of climate change, perhaps the major challenge facing all of us today. Those fifty or so young scientists and students have made, and are still making, a remarkable contribution. This is effectively demonstrated in chapter 10, which recounts many of their achievements.

I have dedicated the book jointly to Cuchlaine King and the late Olav Løken. Olav's leadership in the field was the single most critical assist I could wish for during the later years when administrative and political pressures in Ottawa limited my own Baffin Island involvement.

Many thanks for the highly supportive foreword written by the Honourable Peter Adams, Arctic field researcher, professor, and former director of the McGill Sub-Arctic Research Laboratory.

Dave Andrews, of Digital Art and Restoration in Ottawa, undertook the skillful adaption of the photographs for publication, and Barbara Gordon of Ottawa provided vital cartographic support. I am especially grateful for the editorial care and strong sympathetic support provided by Elisa Hart, editor of the AINA Northern Lights series. Without the care and support of Peter Enman, editor at University of Calgary Press, this book may never have seen the light of day. Alison Jacques impressed me with her meticulous care as copyeditor, for which profound thanks. Special thanks are due to Melina Cusano for turning the process of layout into an art form. Finally, Pauline Ives has provided total support for the past sixty-three years: companionship in isolated parts of Labrador-Ungava, care of small children while I squandered family time in Baffin Island, and so much more.

Jack D. Ives, Ottawa
April 15, 2015

INTRODUCTION

The Context of Geographical Research in Canada's Arctic in the 1960s

This book is a narrative of field research and adventure in Baffin Island, Arctic Canada, a half century ago. The Distant Early Warning Line (DEW Line), a major Cold War–era collaboration between Canada and the United States, had just begun to function through a string of manned radar installations from Alaska across the Canadian Arctic to Greenland. It was essentially an early warning system for detection of a possible airborne atomic attack coming across the Arctic from the Soviet Union. Yet the existence of its physical facilities in some of the remotest areas of the North had an interesting side effect for some Canadian scientists. While still somewhat secret, the DEW Line was radically changing access to remote Arctic settlements and facilitating the reconnaissance-scale topographical and geological mapping of Canada's vast and hitherto almost inaccessible northern lands.

The 1950s and 1960s have often been described, in retrospect, as the "golden age" of Canadian federal government research in the North. There were remarkable opportunities for what may be termed "curiosity" research, which was especially relevant to the discipline of geography despite its being widely regarded as the insignificant stepchild of several other disciplines. In the vast and little-known Arctic and Subarctic regions, the physical extent of which dominated the entire country, such research or reconnaissance was justified, in part, as a process of getting to know and understand Canada as a whole. The primary exploration of this country, the second largest in the world, had been completed shortly after the Second World War, but the secondary exploration of the North became a national duty. This sense of duty was highly relevant on a personal level for me—a "landed immigrant" from Britain and, since 1960, a Canadian citizen—for I had long nurtured an ambition to undertake research in the Arctic. I knew well that these huge "empty" expanses on the map of Canada were larger than France, the United Kingdom, or a combined Norway and Sweden. For me and many of my contemporaries, the exciting prospect of pioneer reconnaissance in the Arctic lands and seas was a palpable and highly motivating force.

In the autumn of 1960, I accepted a senior position in what was then called the Geographical Branch of the Department of Mines and Technical Surveys in the federal government. The position of assistant director, without any prevarication, should itself have been sufficient inducement for a twenty-eight-year-old immigrant to Canada with adventurous ambitions. However, I was especially attracted by a commitment from the branch director, Dr. Norman L. Nicholson, to support a series of field expeditions to north-central Baffin Island, and I was more than pleased to see that I had captured his imagination with my outline of proposed research in that region.

The broader set of my responsibilities included supervising applied research of a practical nature, ranging from analysis of the length of the shipping season for the St. Lawrence Seaway Authority to a study of the rationalization of the Prairie railroad branch line system and the further development of the National Atlas of Canada. But it was the Arctic field research opportunity that so strongly influenced my decision to join the federal civil service rather than to accept a university faculty position.

A first for Canada in the Arctic

This book is about Baffin Island research in the 1960s: specifically, the efforts to set up a continuing series of research expeditions in the North and, in so doing, to establish Canada's first specific federal glaciological unit and a pattern of continuing multidisciplinary and interdisciplinary research that was unique for its time. The Baffin Island expeditions also were a landmark in what we today call gender equality (and which perhaps we no longer consider as remarkable as it was at the time). In the face of strenuous official opposition, we included women as equal members of the field operations: as senior researchers and permanent staff, as well as undergraduate and graduate field assistants.

Another issue was very important. My experience as a doctoral student at McGill University (1954–1956) and as director of the McGill Sub-Arctic Research Laboratory in central Labrador-Ungava (1957–1960) had made me realize that, at the time,

Canada had a serious shortage of young scholars committed to Northern research.[1] The McGill Lab was heavily dependent on attracting young graduates from the UK and several western European countries to serve as year-round staff. It struck me that if a long-term research operation could be set up in Baffin Island through the work of the Geographical Branch, the recruitment of Canadian undergraduate and graduate students as summer field assistants would provide a useful training opportunity. Subsequently, the increasing number of university geography departments at that time (as part of the rapid creation of new universities across the country) would have a potential source for recruitment of young new faculty who would come with extensive Arctic field experience. The Baffin Island undertaking involved a wide range of research training that had a significant impact on later university activities in the Arctic. Many of the summer student field assistants went on to develop remarkable careers, and I have been fortunate to keep in touch with some of them over almost fifty years. Chapter 10 is devoted to exploring this outgrowth of the Baffin Island expeditions, probably the most important contribution of all.

This worthy aim of bringing bright young researchers into both academia and federal government research was not without its structural challenges. In 1967, the Geographical Branch was disbanded organizationally and its different units assigned to separate areas of the federal government. I served as its last director (1963–1967). A sense of desperate defeat pervaded my thoughts at the time, especially in view of the wide range of successes achieved by the Geographical Branch. But the Baffin Island adventure (it can be described as nothing less) did not fade away with the branch in 1967. Like the proverbial phoenix, it took on a new life centred on the University of Colorado's Institute of Arctic and Alpine Research (INSTAAR), which I directed and reformed between 1967 and 1980. Under the leadership of Professor John Andrews, who had been one of the first of the Baffin Island summer graduate student field assistants, research in Baffin Island and many other parts of the Arctic continued to flourish. After 1967, the INSTAAR progression into Arctic and alpine research expanded, developing an increasingly wider set of objectives and drawing in many individuals and institutions from Canada, the United States, the UK, and several European countries. Olav Løken moved as head of the Glaciology Section (to become the Glaciology Division) into the new Inland Waters Branch, thereby preserving its activities that have persisted until the present day.

Increasing relevance of the Baffin Island research

Today, the ongoing activities are by no means centralized as they were in the 1960s, in a small branch of a government department. With the growing awareness of climate change, those early field observations of a half century ago have taken on a heightened relevance. For instance, one especially intriguing discovery of the early 1960s was George Falconer's collection of apparently dead vegetation that was being exposed by the retreating margins of small ice patches on the plateau north of the Barnes Ice Cap. Falconer (1966) even suggested that some of the plant material may have survived alive for several hundred years.

In the context of today's warming climate, another facet of the research is worth recounting. Fifty years ago, we hypothesized a process of climate cooling. The late 1950s and the 1960s were characterized by falling annual air temperatures in many parts of the northern hemisphere; early results from our 1961 reconnaissance showed that, two to four centuries earlier, Baffin Island had reached a so-called "tipping point" of imminent expansion of snow and ice cover—an "instantaneous glacierization" (Ives, 1957). In other words, the climatic and topographic conditions had led to the rapid formation and extension of long-lying snow and ice cover, even with the relatively small amount of temperature lowering. Were we,

therefore, about to experience in the 1960s a repeat of that earlier glacial encroachment (another Little Ice Age, if not a major ice age)?

A few years later, our findings suggested that about eight thousand years ago, the Laurentide Ice Sheet of the last great ice age had disintegrated catastrophically as the North Atlantic waters entered Hudson Bay. At that time, the inrushing waters from the Atlantic Ocean reached a height of up to three hundred metres above present sea level in the southeast section of the bay (Falconer, Ives, Løken, & Andrews, 1965). This hypothesis has since received substantial scientific support, based on the discovery that a leading mechanism was the outbreak of the immense Glacial Lake Agassiz to the southwest, itself provoked by the progressive and rapid thinning of the Laurentide Ice Sheet. The major remnant of the Laurentide Ice Sheet, centred over Foxe Basin and Baffin Island, thinned and receded over the next eight thousand years (excepting periodic reversals, especially during the period AD 1500–1900, the so-called Little Ice Age) until only a chunk of ice, about 145 kilometres long and more than six hundred metres thick, was left sitting on the north-central Baffin Island plateau: this was the Barnes Ice Cap, a last remnant of the ice ages. We further hypothesized that, if the ice cap could be melted away completely by artificial means, it would not reform under the existing climatic regime, even though somewhat cooler than that of today. This is because the underlying bedrock surface would not be high enough to reach the theoretical "permanent snowline" (glaciation level). Thus, each successive winter's snowfall would melt and drain off the Baffin north-central plateau during most of the following summers, as it did from even the highest parts of the Barnes Ice Cap on several occasions in the 1960s; no permanent ice would accumulate.

Another facet of fieldwork emerged from research by Patrick J. (Pat) Webber, then a Queen's University graduate student botanist/ecologist. He established almost one hundred precisely mapped and recorded vegetation quadrats around the northwestern margins of the Barnes Ice Cap. Subsequent visits to many of the field sites have demonstrated a significant increase in the number of plant species, their rate of growth, and total ground cover. This is attributed to a combination of a longer elapsed time following continued retreat of the ice cap (that is, the last fifty years) and the current climate warming. Pat, in collaboration with John T. Andrews, also applied the techniques of lichenometry developed by his mentor, Roland Beschel of Queen's University, to expand his new method for dating rock surfaces that were exposed progressively by the retreating ice cover over the last thousand years or so.

A most significant overseas recruit for the glaciological work was Gunnar Østrem, a Norwegian. Gunnar not only induced us to dig out and fly to Ottawa more than one thousand kilograms of ice from the Barnes Ice Cap, he also introduced to the Canadian Arctic new methods in glacio-hydrology and laid the foundations for a glacier mass balance transect across the Canadian Rockies and Pacific Coast Ranges. This research led to a permanent mass balance record of the Peyto and Place glaciers in Alberta and British Columbia, respectively, maintained to this day as one of Canada's long-term commitments to international glacier mass balance studies.

How this book is organized

The main focus of this book is the logistic, social, and personal aspects of mounting long-term Arctic research at a time when the northern interior of Baffin Island was almost unknown, or unvisited, territory. The chapters and related material constitute a record of the difficulties, the challenges, and the outstanding opportunities of carrying out such a venture from within a branch of the Canadian government in the 1960s. As I hope to attract a wide range of readership and interest to these pages, most of the scientific endeavours are described in general terms, with selected further references provided in the chapter end notes. The final chapter is reserved for a more intensive presentation of the scientific results and a discussion of their impact on subsequent research into the present century.

The first chapter provides a brief geographical introduction to Baffin Island, followed by a short synthesis of previous research. The remaining chapters are arranged in chronological order. Chapter 2 is an account of the pivotal and somewhat chaotic 1961 reconnaissance. It details the uncertainties and difficulties that had to be overcome. The summer of 1961 proved to be a learning experience that laid the foundation for the following field seasons. Chapters 3 through 8 are concerned with activities from 1962 to 1966. The number of personnel ranged from five in 1961 to thirty-eight in 1966, not counting pilots and support engineers. It included Canadian, British, Norwegian, Swedish, Danish, Swiss, and Australian participants. The research disciplines involved climatology, geomorphology, bedrock geology, glaciology, hydrology, plant ecology, animal ecology, geophysics, micro-palaeontology, and oceanography. Eventually, several federal government branches and more than a dozen universities were represented. Chapter 9 relates my transition from Ottawa, via Baffin Island, to the University of Colorado. Chapter 10, as mentioned above, is devoted to an outline of the subsequent career development and achievements of many of the expedition participants, both research staff and student assistants.

The final chapter is intended as a somewhat separate entity, to be passed over by readers who are not especially concerned with the give and take of progress in the natural sciences. As chapter 11 shows, a great amount of new information was acquired and a series of hypotheses developed. Many of the results, while relevant today, have been challenged by subsequent fieldwork, and significant adjustments to some of the original interpretations have been made. Nevertheless, the Baffin Island research of the 1960s provided the basis for the extensive amount of field and laboratory work that has since been undertaken and that continues today.

The large number of Hasselblad colour photos have been included because of their unique contribution as a record of the inspiring Baffin Island landscape as well as their value as documentation of the glaciers, ice caps, and associated landforms as they existed a half century ago, at a time when few were thinking about the potential impacts of climate warming.

My colleague and close friend Olav Løken provided the essential field leadership for the final years of the operation in Baffin, during which my direct involvement was limited by the responsibilities of branch director and the unremitting, yet vain, struggle to preserve the integrity of the Geographical Branch. After the disbanding of the branch in 1967, Olav served as head of the Glaciology Division in the newly formed Inland Waters Branch, Department of Energy, Mines, and Resources, thereby ensuring the survival of a formally designated glaciology unit within the Canadian federal government system.[2]

This account has been written with a strong personal perspective. Had it been prepared by any of the other principal participants, it would likely have a somewhat different flavour—inevitable in view of the great range of activities and the involvement of so many. Because of the inherent likelihood of bias, I have taken extra care to obtain comments, additions, and criticisms from many of those who shared the experience with me. My debt to them is detailed in the acknowledgements and in the relevant sections of the text.

NOTE ON PLACE NAMES

In most of this text, the place names given are those that were in use during the 1960s. For several of the more significant places, current official names are also indicated, as follows: Frobisher Bay (Iqaluit).

Fig. 1: Experiencing the majesty of these remote mountains and glaciers was always an important, if subsidiary, personal objective. This view was obtained approximately midway between the Inugsuin base camp and the outer Baffin Bay coast, north of Itirbilung Fiord, Henry Kater Peninsula, and a wilderness of nameless mountains and glaciers. (Photo: July 1966)

CHAPTER I

BAFFIN ISLAND

The Place and the Research

This chapter provides a framework for the main narrative that follows. It includes a brief overview of the physical geography, notes on early exploration, and an outline of the relatively modest research accomplished prior to the Geographical Branch enterprise. It also outlines the early progress in naming the island's geographical features, bearing in mind that this "anglo" nomenclature usually bypassed the Inuit place names that were based on an oral tradition and were at the time somewhat inaccessible to us. The chapter concludes with a summary of the initial field research objectives.

General geography and early history

Baffin Island (Map 1)—the fifth largest island in the world—is Canada's most extensive island. From the southern tip, it extends north-northwest more than 1,600 kilometres between latitudes 62° and 74° north. Its narrow "waist" stretches about 320 kilometres across from Baffin Bay to Foxe Basin. Lying well north of the treeline, the island is fully Arctic; the Arctic Circle cuts across its southern third. In the 1960s, its population was a scant few thousand, principally Inuit living in small coastal settlements of several hundred people, many of whom still travelled to spring and summer hunting and fishing encampments. Frobisher Bay (Iqaluit) was the main administrative centre. Today it is the capital of Canada's newest territory, Nunavut, but in the 1960s it was still a part of the Northwest Territories.

From a topographical point of view, Baffin Island can be described as the mirror image of the Scandinavian peninsula (Norway and Sweden), although situated about three degrees of latitude farther north. The very significant differences in climate, vegetation, and general habitability between the two land masses, however, are a reflection of their respective locations on opposite sides of the North Atlantic. On the Scandinavian side, the North Atlantic Drift, an extension of the warm Gulf Stream, keeps the far north of Norway (to 71° N) open all winter. However, the spectacular mountain and fiord landscape of Norway's northwest coast is rivalled by Baffin Island's northeast coast, with its many summits rising precipitously to more than 1,500 metres above sea level. (Fig. 1)

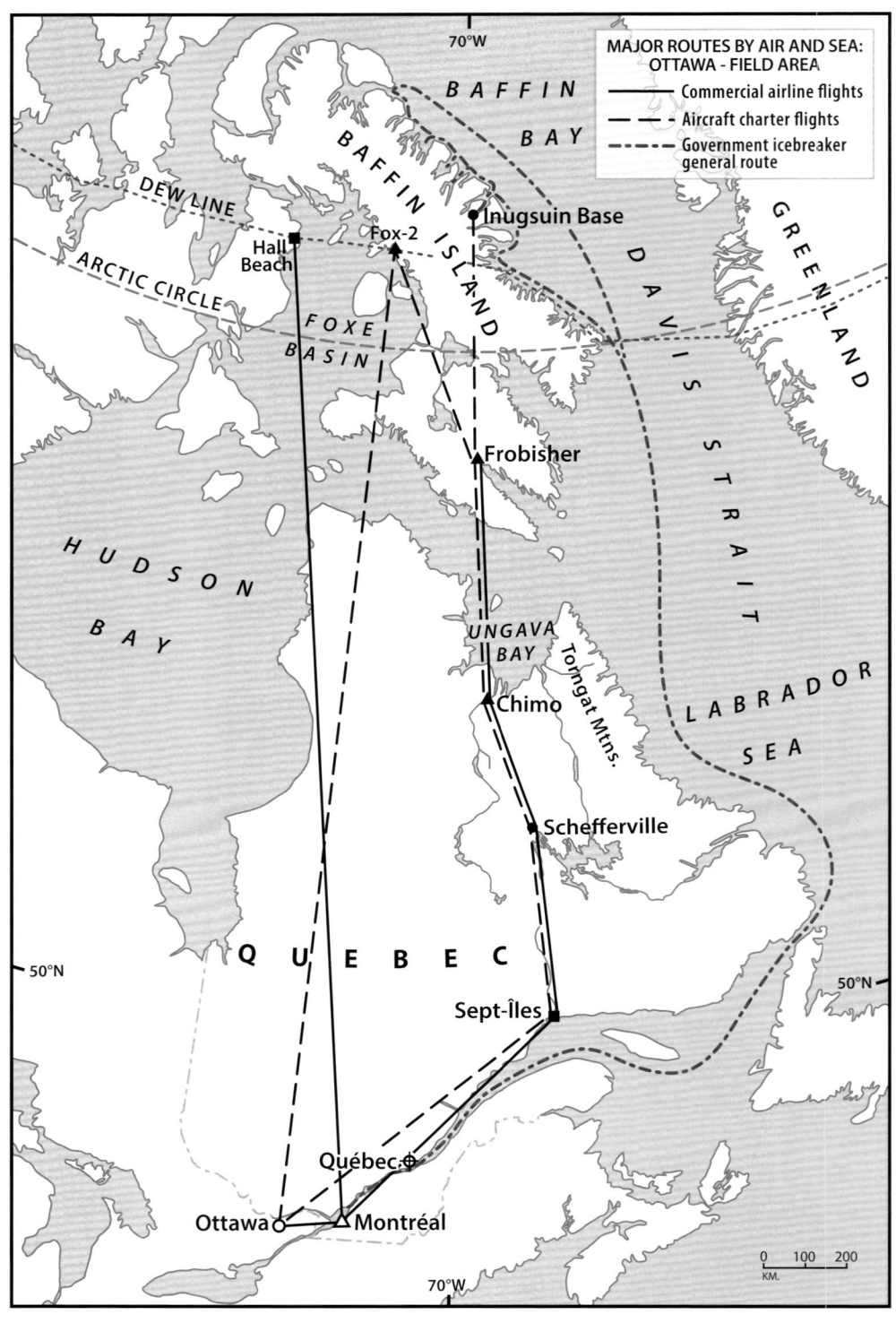

MAP 1:
Part of Eastern Canada showing the location of Baffin Island, the Arctic Circle, and the DEW Line, as well as the main connecting routes between Ottawa and the field area contact sites, close to latitude 70° north, and the generalized icebreaker route travelled in 1966 and 1967 (see inset key).

MAJOR ROUTES BY AIR AND SEA: OTTAWA - FIELD AREA
— Commercial airline flights
– – Aircraft charter flights
–·–·– Government icebreaker general route

BAFFIN BAY

DEW LINE

ARCTIC CIRCLE

Hall Beach

Fox-2

Inugsuin Base

BAFFIN ISLAND

GREENLAND

FOXE BASIN

Frobisher

DAVIS STRAIT

HUDSON BAY

UNGAVA BAY

Torngat Mtns.

Chimo

LABRADOR SEA

Schefferville

QUEBEC

50°N

Sept-Îles

50°N

Québec

Ottawa

Montréal

0   100   200
KM.

70°W

Numerous glaciers and ice caps mantle the coastal mountains of Baffin Island from which the land slopes down gradually southwestward to a broad, gently rolling inland plateau. The surface of the plateau itself continues to fall southwestward to the extensive lowlands and low islands that form the eastern margins of the often ice-choked Foxe Basin. An unusual feature of the north-central interior is the Barnes Ice Cap, more than 600 metres thick and 145 kilometres long, which was suspected to be a relic of the Laurentide Ice Sheet of the last ice age (Baird & Ward, 1952; Goldthwait, 1951).

In 1961, Baffin Island could be classed as little-known despite its immense size. Major islands in Foxe Basin, such as Prince Charles and Air Force islands, and sections of the west coast were discovered or mapped only as recently as 1948, during RCAF air photography operations mounted after the Second World War. The island's interior had been traversed rarely, although for centuries Inuit caribou-hunting parties had crossed between Pond Inlet and northeastern Foxe Basin in late winter and early spring.[1]

Explorers and place names

The southwest coast of Greenland was settled more than a thousand years ago by Icelandic farmers and seafarers (Vikings) who were undoubtedly aware of southeastern Baffin Island.[2] However, it was not until the efforts of adventurous navigators of Elizabethan England, in search of a commercial route north of the North American landmass to the riches of the Orient, that Baffin Island became better known. Hence the names of those early explorers are implanted on modern maps of the island and its surrounding seas: William Baffin, John Bylot, Henry Hudson, Luke Foxe, John Davis, and Martin Frobisher are among the most notable. Long after Baffin and his navigator, Bylot, penetrated to the island's northern tip in 1616 and reached the entrance to Lancaster Sound,

its northern part was still known as Cockburn Land and was believed to be a separate island.[3]

The next sequence of European place names[4] can be largely attributed to the activities of archaeologists and anthropologists in the late nineteenth century and the first half of the twentieth; thus, we have features named for Hantsch, Boas, Rowley, Bray, and Soper, as well as several others proposed by the Danish Fifth Thule Expedition of 1921–1924 (Mathiassen, 1933). Geographers and geologists received their share of recognition slightly later, either directly (Baird Peninsula, Longstaff Bluff) or indirectly according to their Scottish or Cambridge university derivations, such as Cambridge Fiord, Clyde Inlet, Buchan Gulf, Scott Inlet, the Bruce Mountains, Rannoch Arm, Royal Society Fiord (Wordie, 1938).[5] The Scottish whalers of the nineteenth century penetrated many of the eastern fiords, although their discoveries were usually kept as commercial secrets. Nevertheless, later admirers of their exploits immortalized Peter Pond (Pond Inlet, strait, and settlement) and William Penny (Penny Ice Cap). It was the Wordie Expedition of 1937, however, that supported the climbing exploit that led to the first distant sighting from a mountaintop of what was later named the Barnes Ice Cap. In 1950, P. D. Baird, as leader of the Arctic Institute of North America (AINA) expedition to the ice cap, proposed the name in honour of McGill University Professor of Physics Howard T. Barnes.

One of the arduous early crossings of the northern interior was achieved by Graham Rowley, by dog sled, during the British-Canadian Arctic Expedition of the late 1930s. This led to the naming of both the Rowley River and Rowley Island in northeastern Foxe Basin.

In 1940, the Reverend Maurice Flint (Flint Lake) made a winter crossing south of the Barnes Ice Cap from Piling Bay to Clyde Inlet, and the following year, Canon J. H. Turner (Turner Glacier) made the same journey but in the opposite direction. Then, in 1945, RCMP officers Webster and Kyak made a spring crossing using a similar route. None of these

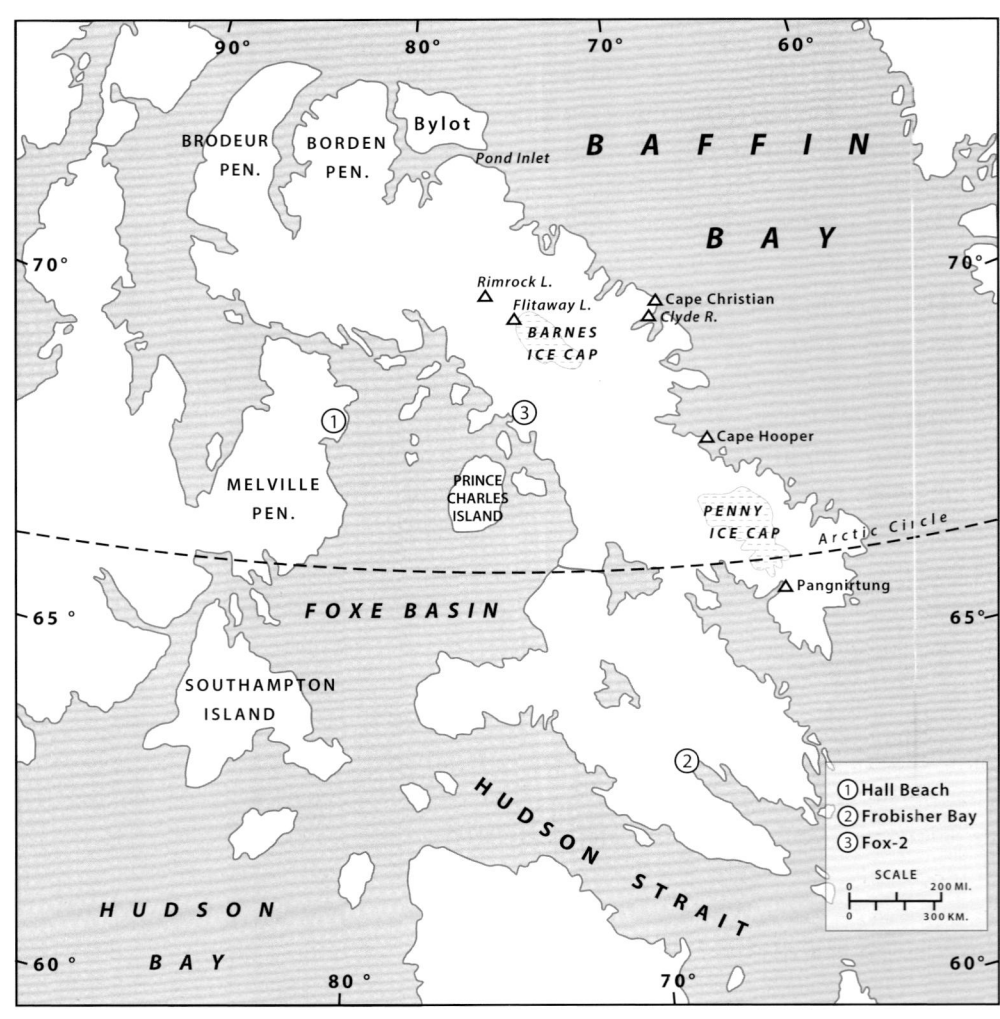

MAP 2: Baffin Island and its surrounding seas, together with selected settlements, Rimrock and Flitaway lakes, and the Barnes Ice Cap.

journeys, however, resulted in any significant new topographic information.

The most recent set of place names approved before 1961 resulted from the highly successful AINA glaciological expeditions led by Colonel P. D. Baird, in 1950 to the Barnes Ice Cap and Clyde Inlet and three years later to the Penny Ice Cap and Pangnirtung. This nomenclature included (from 1950) the Barnes Ice Cap, Generator Lake, Ayr Lake, Eglinton Tower, Broad Peak, Sam Ford Fiord, Walker Arm, Inugsuin Fiord, and (from 1953) Mount Asgard,

Mount Fleming, Mount Battle, Coronation Fiord and Glacier, Maktak Fiord, and Highway Glacier. The 1950 and 1953 expeditions also facilitated the first serious mountaineering in Baffin Island, including the first ascent of Mount Asgard by the Swiss members of the 1953 expedition: Jürg Marmet, Hans (J.) Röthlisberger; Fritz (F. H.) Schwarzenbach, and Housi (J. R.) Weber.

The 1960–1961 winter preparations for the first Geographical Branch research efforts on Baffin Island quickly brought me face to face with the problem of working in a large interior area that was almost entirely lacking in established place names, although there were probably many Inuit names unknown to us at the time. This was immediately prior to introduction of federal government policy in the late 1960s to give preference to First Nation and Inuit place names.

One approach was to extend existing place names to proximate features. Thus, the large river flowing southwestward into Isortoq Fiord became the River Isortoq, and its long lake expansion, Isortoq Lake. New names were also proposed: Rimrock Lake, River Freshney, Striding River, Lewis Glacier and Lewis River,[6] Flitaway Lake, Separation Lake, Windless Lake, and so on. Our main topographic map sheet became "Cockburn Land." The name "Isortoq" was applied to the 1:250 000 map sheet that centred on Rimrock Lake and the northern section of the Barnes Ice Cap. These names and others were formally submitted to, and accepted by, the Canadian Permanent Committee on Geographical Names after the 1961 field reconnaissance to north-central Baffin Island. Subsequently, many more place names were proposed, especially for the fiord area of the northeast coast, as our fieldwork expanded.

Impact of the DEW Line on Arctic access

One of the major consequences of the Cold War relevant to accessibility throughout the Canadian Arctic was establishment of the DEW Line. This was a series of high-tech radar stations constructed between 1954 and 1957 as a joint Canadian–United States venture. Sophisticated and permanently manned installations were set up at approximately 160-kilometre intervals between Greenland, Baffin Island, the central Arctic mainland, and the north coast of Alaska. These main stations closely followed the 69° N parallel with intermediate facilities (I-sites), creating a continent-wide chain with radar sites at 80-kilometre intervals. Short names were applied according to each sector;

for instance, the eastern sector used the place name from Foxe Basin. (Map 2) Its centre of operations, Hall Beach, just south of Igloolik on the Melville Peninsula, was known as Fox-Main. Extending eastward across Foxe Basin and Baffin Island were Fox-1 (Rowley Island), Fox-2 (situated above Longstaff Bluff and often just called "Longstaff"), and Fox-3 (Dewar Lakes, or Mid-Baffin). Cape Hooper on the outer east coast facing Baffin Bay was an I-site. There were east coast links with other defence systems, such as the U.S. Coast Guard LORAN (long range navigation) site at Cape Christian, a few kilometres northeast of Clyde River settlement.[7] In addition, there was a complex chain of sites connecting with the Mid-Canada Line (at approximately 54° N), and the Pinetree Line close to the Canada-U.S. border. The principal DEW Line sites had airstrips and maintenance facilities capable of servicing large aircraft.

This enormously expensive radar system was conceived as an essential line of defence to provide the North American Air Defense Command (NORAD) with early warning of any possible Soviet airborne attack across the Arctic Ocean. Although it was still more or less secret in 1961, it revolutionized research in the Canadian Arctic by providing an indispensable logistical network. Researchers no longer needed to travel north by ship late one summer, overwinter, and conduct their fieldwork the following summer.

My long-range plans for research in north-central Baffin Island would not have been possible without the DEW Line. Initially, our main dependency was on Fox-2 (Longstaff) and, during later summers, on Fox-3 (Dewar Lakes) and, to a much lesser extent, Cape Hooper. The lateral flights along the DEW Line also provided an important link with Fox-Main and comparatively easy access to commercial flights south from there (Hall Beach airport) to Montreal. Another frequently used route was by chartered aircraft between Fox-2 and Hall Beach (Sanirajak) and Frobisher Bay (Iqaluit), also connecting with commercial airline flights to and from Montreal. (See Map 1.)

Previous studies in physical geography and glaciology

Some of the earliest references to Baffin Island in the context of ice age history can be attributed to Bell (1884) and Low (1906). Their publications were understandably superficial considering the limited access from the sea, principally along the north coast of Hudson Strait, and the much more severe sea ice conditions compared with those prevailing today. As mentioned earlier, the 1937 Cambridge University expedition (Wordie, 1938) made a more detailed reconnaissance, using a chartered Norwegian sealer to penetrate several of the fiords of the northeast coast. Wordie's party determined the height of former sea levels up to 60 metres above present sea level along the outer fiords. They also reported much higher terrace features closer to the fiord heads, up to 375 metres, but these were estimated from shipboard and were not visited. Despite doubt that they were marine terraces (Wordie thought it likely that they were glacial lateral features laid down by long-vanished glaciers, and we later proved this to be correct), Professor Richard F. Flint (Yale University), the doyen of glacial geologists of the time, was quick to assume they represented marine features, and they appeared as such on the first *Glacial Map of North America* (Flint et al., 1945).

The British-Canadian Arctic Expedition of the late 1930s, led by Tom Manning, provided invaluable archaeological and ethnographical data as well as maps of much of the previously unmapped Foxe Basin coastline and islands (Rowley, 2007). However, Baird's long journey of 1938 from Igloolik via Steensby Inlet to Piling Bay (south of the site that would become the Fox-2 DEW Line station) produced little more than topographical notes.

The first serious and sustained fieldwork inland from the coasts came with the 1950 and 1953 AINA glaciological expeditions previously mentioned (Baird & Ward 1952; Ward & Baird, 1954 Weber, Marmet, Röthlisberger, & Schwarzenbach, 2008). Their work, centred on the southern dome of the Barnes Ice Cap and on the Penny Ice Cap, also included botanical (Schwarzenbach, 2011) and zoological (Watson, 2011) research. One of the more significant conclusions was the recognition of the Barnes Ice Cap as a geophysically new type of glacier—one where all the previous winter's snow frequently melted and where glacial mass was maintained by refreezing of snowmelt onto the cold underlying ice. Ward and Baird (1952) proposed the term "Baffin-Type" for glaciers that received their nourishment solely from the refreezing of the snow meltwater, although they were partially anticipated by Schytt (1949) from research on the glaciers of the Kebnekaise area of Arctic Sweden. Richard Goldthwait was the senior Pleistocene geologist on the 1950 expedition. He observed sets of end moraines extending between the heads of Sam Ford and Clyde fiords (Goldthwait, 1951), although he was not able to gauge their massive extension from west of the Penny Ice Cap to the far northwest of the island. Nevertheless, he generously gave me access to his field notes and invited me to Columbus, Ohio, for discussions during the 1960–1961 winter.[8]

John Mercer's doctoral dissertation, presented to McGill University in 1954, added valuable information on the glacial history of southernmost Baffin Island, although his field area was somewhat peripheral to that of the Geographical Branch and more than one thousand kilometres farther south. Like Flint, he also had become interested in very high terraces along the outer coast of Frobisher Bay. He assumed that they indicated former sea levels up to 435 metres above present sea level (Mercer, 1956). Furthermore, Mercer had worked with me in Ottawa during the 1956–1957 winter, and we had had many discussions about Baffin Island, especially the north-central interior. This influenced my determination to go there four years later.

In 1955, Robert Blackadar, with a Geological Survey of Canada (GSC) bedrock reconnaissance mapping party, recorded glacial features around Admiralty Inlet in the far northwest and introduced evidence for a late ice age flow of ice from the north-west across Steensby Inlet (Blackadar, 1958). In 1960, Bruce Craig and John Fyles, also of the GSC, published a summary paper on available information concerning the glaciation and deglaciation of the Canadian Arctic (Craig & Fyles, 1960). Their accompanying maps emphasized the scarcity of field data available to them, and they made the tentative assumption that, during the last ice age, a Baffin Island–Ellesmere Island glacier complex functioned independently, or semi-independently, of the main Laurentide Ice Sheet. Later research would show this to be incorrect.

Envisioning new field research on Baffin Island

The glimmerings of a plan for fieldwork in north-central Baffin Island had emerged from my first cursory air photo interpretation while working on preparation of the *Arctic Pilot* during the 1956–1957 winter in Ottawa (Canadian Hydrographic Service, 1959). Of particular interest were systems of massive end moraine that could be traced for hundreds of kilometres parallel to the heads of the fiords, extending down the length of the fiords to Baffin Bay. Next, there was a curious light- and dark-toned pattern widely spread across the interior plateau. Finally, what appeared as the shorelines of former glacially dammed lakes could be traced in several of the major valleys north and west of the Barnes Ice Cap with innumerable linear features perpendicular to them (see Figs. 2 and 3).

The potential importance of the four sets of features had been reinforced by the results of five years of fieldwork across Labrador-Ungava in the late 1950s and my increasing knowledge of the more detailed research in Scandinavia (Hoppe, 1952, 1959; Ives, 2010). The work of Professor Gunnar Hoppe (Stockholm University) in northern Scandinavia had indicated a major reversal of the classic assumption that large lakes had been dammed by a remnant mass of

ice located over the Gulf of Bothnia and had overflowed across the Norwegian frontier into the Atlantic fiords. This phenomenon was seen as evidence of ice sheet and land interrelationships during the final phases of the Fenno-Scandinavian Ice Sheet of the last ice age. Hoppe had effectively demonstrated in the field in 1960, during an international symposium, that there had been no large ice-dammed lakes, but rather, merely small lakes in the tributary valleys dammed laterally by large glaciers in the main valleys that were flowing eastward to the Gulf of Bothnia. This led to the conclusion that the final flow of glacier ice toward the end of the last ice age was from the Norwegian-Swedish border mountains eastward into the Gulf of Bothnia, not the reverse. My work in the Torngat Mountains and along the George River in Labrador-Ungava in the 1950s demonstrated an apparently contradictory situation—very large lakes in the main valleys spilling across the watershed into the Labrador Sea with the remnant ice sheet located to the west, or inland, of the Labrador coastal mountains (Ives, 1960b, 2010).

Baird (1950) and Goldthwait (1951) had speculated that the Barnes Ice Cap might be a relic of the Laurentide Ice Sheet.[9] In the late 1950s, when I began reconnaissance mapping of the outlines of the final disappearance of the Laurentide Ice Sheet in central Labrador-Ungava, the map of Baffin Island appeared as its present-day analogy. The Barnes Ice Cap, located anomalously on the plateau and damming glacial lakes along its eastern margins, appeared as a rough parallel, thousands of years later, to the last remnant of the ice sheet in the centre of Labrador-Ungava, damming glacial lakes along its northeastern margins (Ives, 1960b).

Already, the work in the Torngat Mountains of northern Labrador (see also Løken, 1960, 1960, 1962) had challenged the prevailing paradigm, established by Professor Richard Flint, on the origins and eventual disappearance of the Laurentide Ice Sheet during the ice ages. Extending the fieldwork to Baffin

Rimrock Lake

Fig. 2: Vertical view of ice-bound Rimrock Lake. The small circle marks base camp 1 (1961). This photograph shows light and dark surface tones, the exploration of which was one of the four major objectives of the 1961 reconnaissance. The Glacial Lake Lewis shoreline, the second objective, is conspicuous on the hillslope just northeast of the camp as it forms a boundary between light- and dark-toned ground (marked by small arrow). Cross-valley moraines, the third reconnaissance objective, can be seen close to the lower margin, left of centre. The large arrow within the lake points north. (Source: Modified from Ives, 1962, Fig. 4).

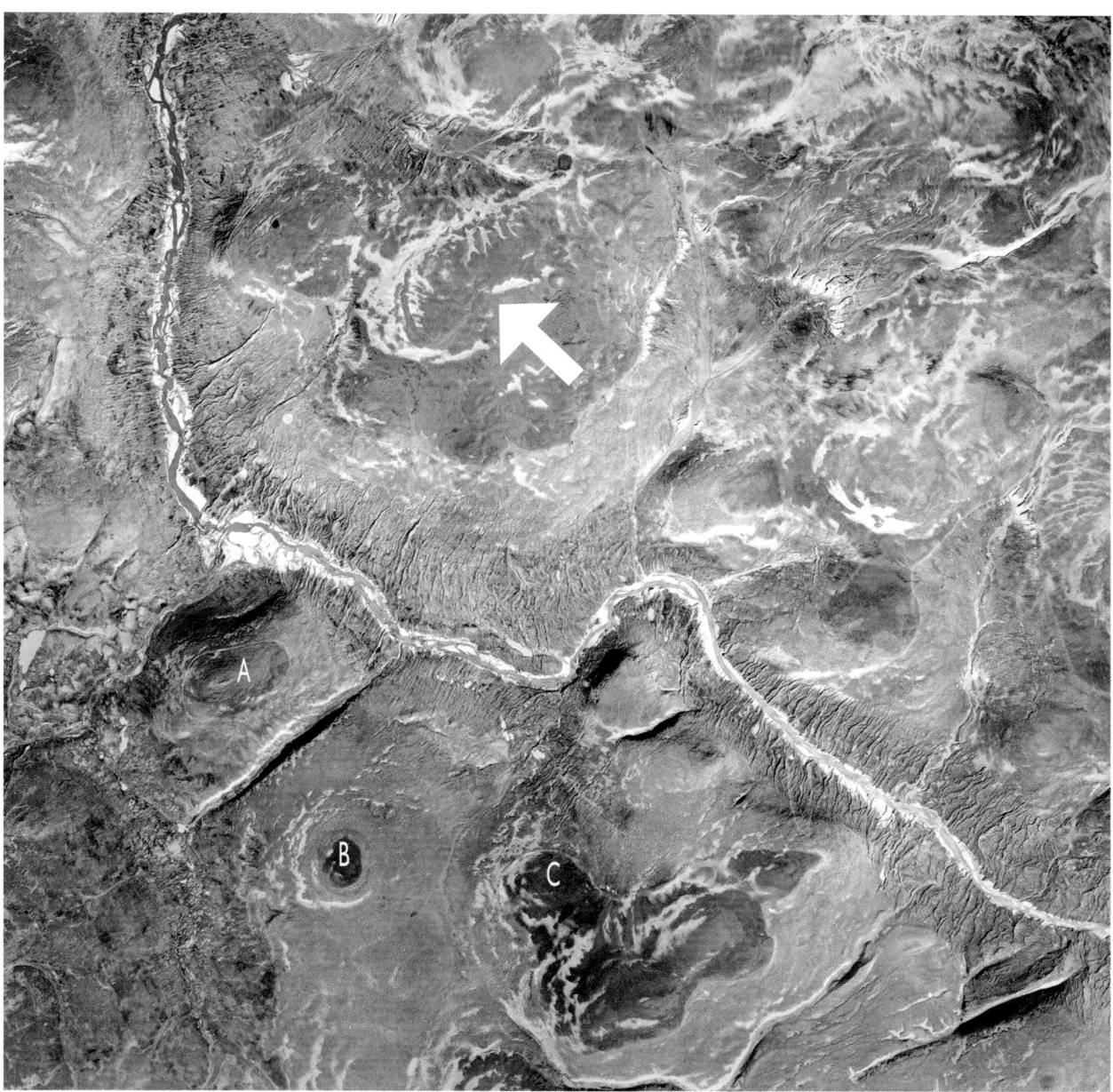

Fig. 3: Vertical view of midsection of the Isortoq valley displaying hundreds of cross-valley moraines trending perpendicular to the river. The Glacial Lake Lewis shoreline is also conspicuous. Note the small former islands—marked A, B, and C—that had risen above the lake surface when it existed. The large arrow points north. (Source: Modified from Ives, 1962, Fig. 3).

Island presumably would strengthen the challenge and greatly widen the scope of the research.

The Labrador-Ungava studies, however, had been conducted on a shoestring budget. Such a limited approach for the far more logistically demanding north-central Baffin Island would be impractical. My new appointment as assistant director of the Geographical Branch in 1960, therefore, was timely in terms of opening access to a much more generous level of funding.

Outline of a field research plan

From this account of the previous, limited, Baffin Island research it can be seen that during the 1960–1961 winter any field plan to evolve would be heavily influenced by the earlier research undertaken in Labrador-Ungava from the McGill Sub-Arctic Research Laboratory and the much more extensive work undertaken by Swedish and Norwegian scholars in Scandinavia. There would also be a significant dependency on study of the air photographs of Baffin Island. Nine separate targets had already been identified and put forward as part of my proposals to Dr. Nicholson in 1960. The bedrock of the agreement was the 1961 reconnaissance (see chapter 2). Even the slim Geographical Branch budget could be stretched to accommodate such an exploratory enterprise. I had been clearly informed that the scale of the long-range plan would depend both on the outcome of the proposed 1961 reconnaissance and on my success in using it to persuade the department for a substantial increase in the branch budget. It transpired that Dr. W. E. van Steenburgh[10] became the key figure, and in retrospect, it seemed that he had personally adopted my scientific aspirations. This will become apparent in the chapters ahead.

THE NINE KEY TARGETS FOR PROPOSED BAFFIN ISLAND RESEARCH

1. To trace, as far as possible, the final phase of the last ice age in the Eastern Arctic setting of north-central Baffin Island. One overriding question was how far the Labrador-Ungava glacial and topographical analogy would apply, given the assumed difference in timing of thousands of years.

2. To determine whether the reconnaissance knowledge of the extensive moraine system that appeared to run parallel to the fiord heads of the northeast coast fitted into this broad analogy.

3. To map the extent of the prominent shorelines of what could be assumed to represent the margins of former ice-dammed lakes, to measure whether or not they were horizontal, and to determine the extent to which they would reflect the withdrawal of the late-glacial ice sheet.

4. To survey the heights of the uppermost marine shore features at several localities along the southwest coast fronting Foxe Basin and, if possible, collect marine mollusc shells (seashells) for radiocarbon dating.

5. To use any field evidence collected under items 3 and 4 for determination of interrelations between a wasting remnant of the Laurentide Ice Sheet, entry of salt water into an emerging Foxe Basin, and the pattern of isostatic uplift of the land.

6. To determine the causes of the remarkable pattern of light and dark tones on the landscape as seen on the air photographs characterizing wide areas north and east of the Barnes Ice Cap.

7. To study some of the hundreds of linear land-forms that trend across the main valley floors in the same general area.

8. To examine recent fluctuations in the margins of the northern part of the Barnes Ice Cap.

9. To supplement all of the above by study of a cross-section of the "waist" of Baffin Island south of the Barnes Ice Cap.

As the area chosen for the 1961 reconnaissance was so remote and hitherto barely investigated, it was assumed that additional research topics would be recognized. Nevertheless, it should have been apparent that my ambitions could have been over-reaching in the extreme. Much would depend on good fortune, even the prior selections of principal field research localities in such a large and essentially unknown region.

As it turned out, choice of base camp locations proved largely successful, although a succession of near-disasters with a single-engined Cessna floatplane could have spelled failure (as will be detailed in chapter 2). Regardless, the large lake located some forty kilometres north-by-west of the northern end of the Barnes Ice Cap (Rimrock Lake: Figs. 2 and 3, pp. 14, 15) proved a vital and successful starting point. Here the air photos showed distinct glacial lake shorelines in close proximity to a large swath of the light- and dark-toned terrain pattern and innumerable examples of the narrow ridges that trended at right angles to the main valleys. The small lake, dammed by the northwestern margin of the Barnes Ice Cap (Flitaway Lake) was identified as the second critical site. Another area that needed investigation was the lower section of the Rowley River leading to Steensby Inlet.

Here the air photos showed both raised marine shore features and glacial moraine systems. This pattern of landforms could be traced all the way down the Foxe Basin coast to the Fox-2 DEW Line station. Finally, a transect from Fox-2 across the entire waist of the island south of the ice cap appeared worthy of careful study as well as a short visit to a section of the extensive moraine system to the east of the ice cap. It was obvious to me at the time that these field plans were rather ambitious. The area chosen was huge and very little of it had been visited previously. The dates of lake ice breakup, critical for the amount of time we would be able to operate a floatplane, were unknown. Nevertheless, I thought it better to have too much rather than too little to accomplish.

In view of available resources, therefore, I decided to fly into Rimrock Lake by ski-equipped airplane early in the season and then transfer to Flitaway Lake and to Steensby Inlet once open water was available for a small floatplane. Vic Sim, a Geographical Branch colleague with previous Canadian Arctic experience, would tackle the transect south of the ice cap and work on the raised marine shore features north of Fox-2 along the Foxe Basin coast, possibly as far north as Steensby Inlet. Vic and I had spent many hours in the National Air Photo Library during the 1960–1961 winter and felt that we were reasonably well prepared. He decided to take Claude Lamothe as his student assistant. I would take two assistants: John Andrews, who had been with me as a graduate student at the McGill Lab and was completing his master's degree in Montreal, and Peter Hill, a junior member of the branch staff, keen for Arctic experience. There would thus be a field team of five operating for the most part as two semi-independent units, ideally in contact by radio, despite being more than 150 kilometres apart for most of the season.

Fig. 4: Fox-3 DEW Line station, central Baffin Island. The DEW Line provided indispensable logistical support for many research expeditions and field surveys during the 1960s and 1970s across the entire Arctic, from Alaska to Greenland. It was vital to the success of Geographical Branch operations. The thin white line along the horizon is the Barnes Ice Cap (Photo: August 1966).

CHAPTER 2

❧ RECONNAISSANCE 1961 ❧

Learning about Airborne Support

At last we were approaching the site selected for our base camp. The atmosphere in the cockpit was tense. Frank Ross, the pilot of our chartered Wheeler Airlines single-engine de Havilland Otter, was quietly cursing the deteriorating weather ahead. As a rank novice at this game, I felt nervous anticipation mounting as we gradually had to creep below the lowering cloud base. So it was with immense relief when, at 8:55 p.m. on June 15, 1961, Frank gently touched the plane's skis down into the mush of melting snow on the still ice-covered expanse of water I later named Rimrock Lake. We had reached latitude 71°41' N, some forty kilometres north of the Barnes Ice Cap, after a speedy journey by air from Ottawa. Only the day before, we had flown from Montreal, via Fort Chimo and Frobisher, to the DEW Line site at Longstaff Bluff (Fox-2). Now we would find out if the preparations of the previous winter in Ottawa, nearly three thousand kilometres to the south, had been adequate. This was my first major venture with the Geographical Branch, Ottawa—an examination of the landscape, glacial history, and glaciology of the interior of this little-known Arctic island.

Peter Hill, a junior member of the Geographical Branch staff, and John Andrews, who had spent the 1959–1960 academic year with me as a graduate assistant at the McGill Sub-Arctic Research Lab in central Labrador-Ungava, made up the "northern party." Peter, although he had no previous field experience, while an undergraduate at McMaster University had developed an enthusiasm for the Arctic from attending lectures by Dr. Hugh Thompson.[1] John was my obvious first choice as summer graduate student assistant because of his highly relevant experience on the Labrador coast the previous year and his 1959–1960 winter at the McGill Lab. Dr. Victor (Vic) Sim, also of the Geographical Branch staff, who had experience in the Canadian Arctic, and his summer graduate assistant, Claude Lamothe, constituted the "southern party." They were to work across the waist of Baffin Island south of the Barnes Ice Cap from bases at the DEW Line stations of Longstaff Bluff (Fox-2) and Dewar Lakes (Fox-3). We had planned a three-month reconnaissance to be followed by a series of larger parties in successive years if this first summer proved successful.

Frank taxied through the wet snow and standing water as close to the southern shore of the lake as he thought wise; a moat of open water seemed to follow the lake's entire perimeter and we were concerned that the thick lake ice might be afloat. Frank and I off-loaded two 45-gallon drums of aviation fuel, tents, and an assortment of crates containing food. Within fifteen minutes, we were taxiing for takeoff in rapidly deteriorating weather, returning to Longstaff Bluff for the second load, which would include John and Peter and the rest of the equipment. But that second trip was not to be. Fog was rapidly enveloping the Foxe Basin coast. An optimistic start to the season in terms of reaching our chosen base camp was to be set back several days.

Air photo studies and origins of the Baffin project

The notion of a venture into the interior of north-central Baffin Island had been maturing in my mind for several years. During the 1956–1957 winter, I had been a member of a small research team of AINA, working under contract with the Canadian Hydrographic Service. Our task had been to produce a manuscript for a new edition of the *Pilot of Arctic Canada* (Canadian Hydrographic Service, 1959). The work included extensive air photo interpretation of all the Canadian Arctic coastal characteristics. I had been allotted Baffin Island (amongst other areas) by our team leader, retired Royal Navy Captain R. M. Southern.[2] In light of my strong personal interest in the history of glaciation, this tempted me to make occasional photographic excursions inland from the coastline, while dodging the disciplinary eye of Capt. Southern.

In the process, I spotted three sets of intriguing features on the air photos. The first was a massive set of glacial moraines extending more than eight hundred kilometres; they were situated inland of, and parallel to, the heads of the fiords that cut through the coastal belt of mountains forming the northeastern section of the island. Moraine offshoots streamed down the steep slopes and along the sides of many of the fiords.

The second feature was a pattern of light- and dark-toned areas that extended from the northeastern mountain rim across the inland plateau north of the Barnes Ice Cap; the area lies far to the north of the Arctic treeline, and the pattern could not easily be explained by assuming that it represented different types of vegetation. It appeared to be related to a combination of topography and direction of the prevailing wind. (Fig. 2) From the confines of the National Air Photo Library, I had speculated that the pattern was related to the deposition of wind-blown snow or sand and that it decidedly warranted investigation. One strikingly regular feature of the light-toned area was the linear coincidence with what appeared to be the shoreline of a former ice-dammed lake that extended more than eighty kilometres.

The third set of conspicuous features, north of the Barnes Ice Cap, was a series of hundreds of small linear forms trending perpendicular to the major valleys; they appeared to be confined below the shoreline of the former ice-dammed lake. (Fig. 3) These landforms, when added to the other two sets of features, appeared to offer great potential for preliminary field investigation that was literally staring me in the face, through my stereoscope.

These somewhat incidental interpretations remained intriguing, fixed in the back of my mind, along with a question: How on earth can I get there to take a look? The thoughts kept resurfacing during my three years (1957–1960) as field director of the McGill Lab[3] located on the edge of the new iron ore mining town of Schefferville and close to the geographic centre of the Labrador-Ungava peninsula.

During my time at the McGill Lab, I had begun to formulate a long-term study of the glacial history of the peninsula (Ives, 1960b, 2010). I speculated on the relevance of the tentative air photo interpretation in Baffin Island to an understanding that was arising from the lab's field program in Labrador-Ungava. During the Abisko Symposium of the 1960 International Geographical Congress in northern Scandinavia, I discussed with Professor Gunnar Hoppe, the symposium leader, the apparently similar pattern of glacial features in all three areas. More than ever, I was convinced of the value of familiarization with the recent research by Swedish and Norwegian scholars.

I entered Canadian government service in September 1960, following my visit to Sweden and Norway, and moved to Ottawa with my family. My duties at the Geographical Branch involved much more, of course, than preparations for a series of expeditions to Baffin Island. In practice, Dr. Nicholson wanted me to assist him in a total reassessment of the aims and structure of the branch. This was set in motion shortly after my arrival. However, I did have sufficient time for a reconnaissance-level air photo interpretation of almost all of Baffin Island. I followed up by reading all the scientific literature and finding that it was by no means extensive.

And so the winter of 1960–1961 passed quickly. The Geographical Branch had moved into an impressive new building during the previous spring and summer, adjacent to the GSC and the Surveys and Mapping Branch, which included the National Air Photo Library. I was well placed to spend time inspecting recently acquired air photos of Baffin Island. And with a view to the larger objectives of the planned long-term field research, I encouraged both George Falconer and Benoit Robitaille of the regular branch staff, who were accompanying government icebreakers down the coast of Baffin Island, to make spot landings wherever possible. This was to check for the existence of raised marine features and the marine limit down the entire east coast.

The first half of the battle: Getting there

Now, after the winter's preparation, we were on the threshold of our new enterprise. Peter Hill and Claude Lamothe had preceded John Andrews and me by six days to make sure that all the equipment and food, sent in advance, had been safely transferred to Fox-2. Vic Sim had left two days before us, taking a direct flight from Montreal to Hall Beach (site of Fox-Main station) on the Melville Peninsula and thence eastward along the DEW Line across Foxe Basin to Fox-2. I had chosen the alternative route via Frobisher Bay

(Iqaluit) so that we could pick up our chartered ski-equipped Otter there, fly directly to Fox-2, meet up with the rest of our party, and, we hoped, continue in two or three Otter-loads to Rimrock Lake, about two hundred kilometres farther north. These were highly optimistic logistical plans that came off with remarkable precision, that is, until the second and subsequent loads between Fox-2 and Rimrock Lake.

We were firmly stuck at Fox-2 waiting for the weather. June 16 was a day of fog and cold light rain, forcing us to enjoy the luxuries that the DEW Line provided to their otherwise isolated station crew: table tennis, eight-ball pool, the latest movies on reel-to-reel projectors, and excellent food, plus outstanding camaraderie with the establishment. The station chief proved to be a larger-than-life character. Lou Riccaboni (a former RAF Battle of Britain Spitfire Wing Commander with a handlebar moustache) expressed great interest in our plans and promised full support, saying he would provide his personal attention and would relay our proposed weekly radio messages from our remote field bases to Ottawa.

Another day of light rain and fog followed, with the temperature reaching 5°C. Our pilot, Frank Ross, was becoming decidedly edgy, anticipating landing with skis on the waterlogged snow of our ice-bound lake. When the fog lifted the following morning and the cloud base had risen to about one thousand metres, he was anxious to "give it a try." John and I, with as much extra equipment as could be packed into the Otter, were down on the airstrip (we were using combination wheel-skis) and belted into our seats by mid-morning. Soon we were flying low over the almost totally snowbound land along the western edge of the Barnes Ice Cap. The farther north we progressed, the lower the cloud base until, toward the north end of the ice cap, we could barely make out the general lie of the land. Farther north in the direction we thought would take us to Rimrock Lake, Frank circled for about twenty minutes in an almost complete whiteout, hardly catching even a glimpse

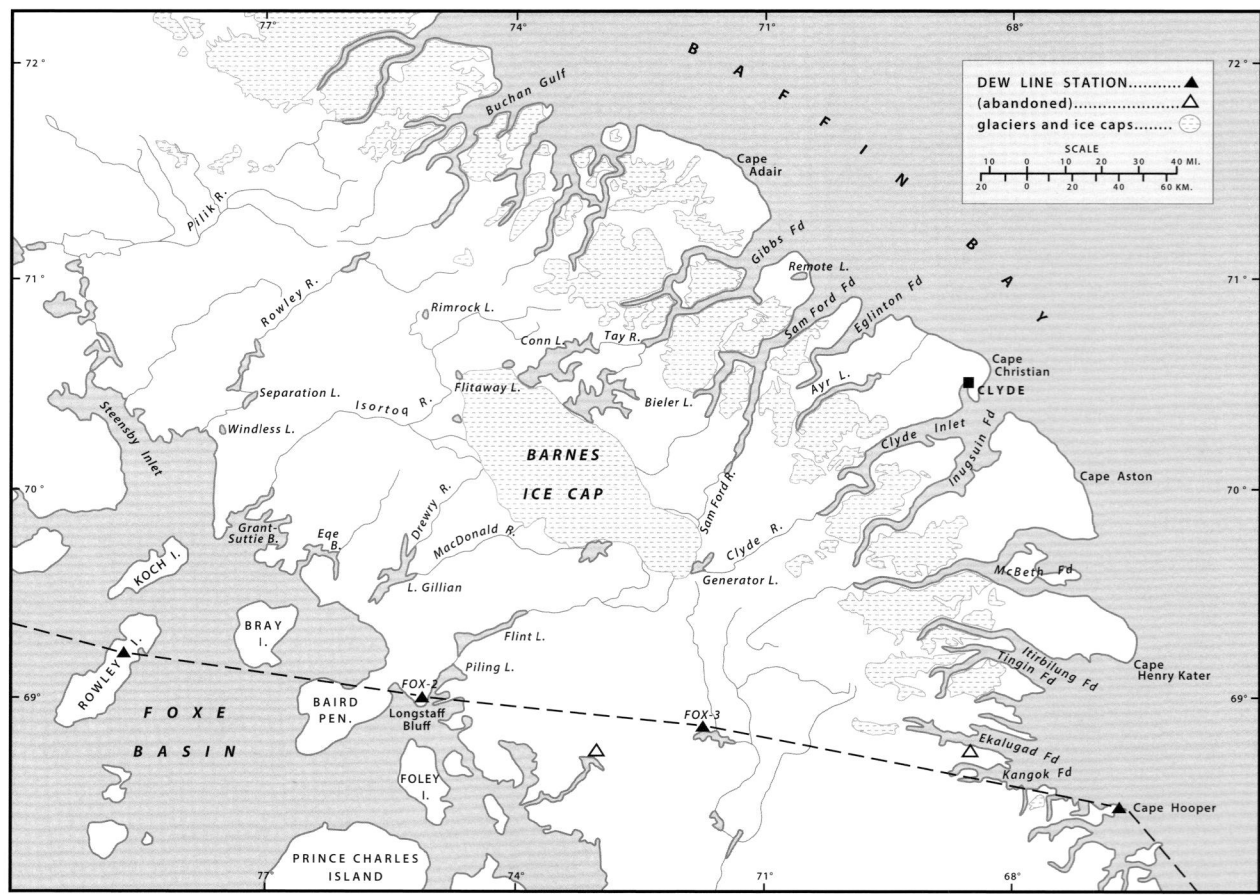

MAP 3: The field area of north-central Baffin Island, emphasizing 1961 activities north and west of the Barnes Ice Cap: the main base camps, at Rimrock Lake, Flitaway Lake, Separation Lake, and Windless Lake.

of the ground. We abandoned hope of a landing and returned to Fox-2.

After another day of forced rest we finally reached our base camp. The morning of June 19 provided clear weather but with a wretched stiff crosswind. With the progression of the rapid melt, I began to worry that I would have to plead with Ottawa for extra funding to charter a helicopter—a very expensive recourse. Vic Sim arrived in the early afternoon, after several days' delay at Hall Beach. Now the entire party was together at Fox-2. Our meeting with Vic was brief, however, as the wind had dropped and Frank was more than anxious to complete his mission to Rimrock Lake and head back to Frobisher Bay.

Frank circled the lake with apprehension. He had anticipated a lot of standing water, but he made a run in toward the shore and reached a spot within a few metres of our cache. Fortunately—and to our surprise—most of the snow had melted, and the standing water had almost entirely drained off. At five o'clock we were hastily unloading. Twenty minutes later, John and I watched the Otter take off again and head back to Fox-2 to pick up Peter and our final load. John and I began a wet-footed slog to take as much of our equipment as possible off the ice and onto a small rocky point at the southeast corner of the lake, a distance of about three hundred metres (Map 3).

The entire area around the lake appeared to be mantled with a thick layer of angular boulders, nearly completely snow-covered and with some of the worst possible terrain for walking. The prospects for a reasonable campsite appeared grim. We searched diligently but failed to find a spot where we could pitch even one of our small mountain tents without laborious effort.

The Otter reappeared at 8:30 p.m. Peter and our remaining equipment were quickly off-loaded. Frank helped us set up the radio antenna and establish our first Fox-2 contact with station chief Lou Riccaboni. Lou's voice nearly split my eardrums, so clear was the radio contact: "Fox-2 receiving you loud and clear. Wizard prang, old chap, all's well and I will relay your weekly message to Ottawa every successive Friday following our sked. Good luck. Over and out." And so our fine scheduled radio contact continued for the next two Fridays—at "0300 hours ZULU" (meaning Greenwich Mean Time in DEW Line lingo).

To our surprise, from July 7 until August 4 there was a month-long total radio blackout. We later learned that the branch head office and my family, come what may, had continued to receive the weekly message from Fox-2 as if I had sent it: *Party all well—Ives.* In this way, Lou had forestalled any unnecessary rescue mission! Because my previous years of fieldwork in Labrador-Ungava had placed my wife and me in isolation without even radio contact with the outside world for much longer periods, I was not perturbed about the radio blackout, nor was John. For Peter, however, on his first Arctic "wilderness" experience, the period of prolonged radio blackout without explanation proved very unnerving. I learned this only later, for there was little indication of his concern at the time. When radio contact was eventually restored, we were told that a period of heavy sunspot activity had disrupted radio communications in a wide area around the North Magnetic Pole.

The Rimrock Lake experience

As we settled into our camp at Rimrock, the infinite silence of the Arctic descended upon us, and there was a pronounced sense of aloneness. A pause for coffee after the Otter's departing roar had faded far to the south gave us our first chance to contemplate our solitude. We gazed upon a predominantly white landscape. Gently rolling hills were set around the seemingly immense expanse of ice-covered water that was Rimrock Lake. Steep faces of the larger boulders closest to us projected darkly through the snow. On the more distant hills, dark outcrops of steeply sloping rock provided some relief to the all-pervading whiteness that persisted for twenty-four hours a day. A narrow moat of open water traced the margins of the lake for as far as we could see. Fortunately, this moat was superimposed on still-thick lake ice attached to the shore, so we had no problem in getting ourselves and our equipment off the lake. Our only discomforts were wet thighs and very cold toes. In the low sun, we could easily make out the apparently horizontal former lake shoreline above the present lake level approximately five hundred metres asl, as determined from the Otter's altimeter. We could rest assured that there was something for our investigative curiosity in the wild landscape of boulders, snow, and water.

It was time to set up a tent and roll out sleeping bags on this mattress of boulders. The big tent (a so-called Jutland building with a light metal frame supporting a strong canvas cover) was pitched and camp cots set up on the uneven floor. With these preparations completed, we turned our attention to dragging more of our essential equipment off the lake ice. We had realized that this would be a problem on our first Otter landing and so had asked Frank, our pilot, to pick up two large wooden floats, promised by Lou over the radio from Fox-2. They served a dual purpose—as crude sledges and as a floor for our main tent. Pulling the loaded sledges was very hard going at first, but as midnight approached and the surface began to freeze, it became much easier. So we worked like stevedores until after three in the morning when we collapsed into our sleeping bags for a very rocky "night" of sound sleep.

We awoke late morning to an empty blue sky and a blinding white landscape. The day was spent dragging equipment off the lake and filling canvas buckets with gravel and mud dug out from scattered frost boils that had broken through the boulder field around the campsite. This finer material was used to fill spaces between the boulders that composed our tent floor. The improvement was noticeable once our improvised sleds were converted to tent flooring.

These activities continued for another day. Finally, we wrestled the five forty-five-gallon drums of aviation fuel to the edge of the lake and lashed them together with a length of thick rope that was looped around a large boulder in the hope that this would prevent their escape into open water when the lake ice broke up. By late afternoon on June 21, our base camp was quite liveable. But forty-eight hours of constant sunshine had melted off much of the snowy mantle and, regrettably, had caused uncomfortable sunburn under chins and in nostrils, where we had neglected to apply adequate sunblock.

For the time being, we were restricted to the general vicinity of Rimrock Lake. That would be the situation at least until our chartered Cessna from Bradley Air Services arrived from Ottawa. But was the Cessna actually on its way? We had no way of knowing. We had been without radio contact with Fox-2 for a whole month prior to August 4. On that day we had learned, to our dismay, that the Cessna seemed to have just disappeared somewhere between Ottawa and Baffin Island. Meanwhile, Rimrock Lake had been showing sufficient open water even by July 30 to permit a floatplane to land, so the unaccountable silence and long delay were hard to understand. We had completed a significant amount of fieldwork in the area around Rimrock Lake that was accessible by foot, but with the delay of the floatplane we were

losing precious time for inspection of the more distant sites essential to our reconnaissance.

During the early period following establishment of the base camp, our mobility had been severely limited. Snow on top of angular boulders was not only uncomfortable but dangerous for day hiking and backpacking. This was the only time in my life during which I suffered from blistered insteps—the price of constantly balancing on the sharp crests of boulders for long distances. Despite this early difficulty, we identified and precisely surveyed a prominent former lake shoreline 505 metres asl[4] and slightly above the level of Rimrock Lake.

With our new automatic Zeiss telescopic level, we established three survey points on sections of the lake shoreline so that they formed the corners of an equilateral triangle with sides fifteen kilometres long. (Map 4) The result showed fairly conclusively that the shoreline of the former glacial lake was horizontal and thus had drained away in the not too distant past. In other words, no detectable isostatic tilt had occurred, implying that the lake had drained hundreds rather than thousands of years ago. I named the feature Glacial Lake Lewis, also for Vaughan Lewis. (Fig. 5) Furthermore, as the snow continued to disappear, and as sunny skies alternated with rain and fog, it became apparent, as I had suspected the previous winter from review of the air photos, that the Glacial Lake Lewis shoreline would be the focal object of our studies. There was a distinct division between large sections of the dark-toned areas (above it) and uninterrupted light-toned areas (below). The explanation became obvious once it was possible to examine the phenomenon on the ground. It resulted from the size (diameter) and degree of cover of rock lichens (principally *Rhizocarpon geographicum* species). Below the shoreline, the lichen cover was minimal and the individual lichen diameters were no more than a few millimetres, thus giving a very light appearance due to the large proportion of exposed fresh rock surface. However, above the shoreline there was an irregular spread of light- and dark-toned

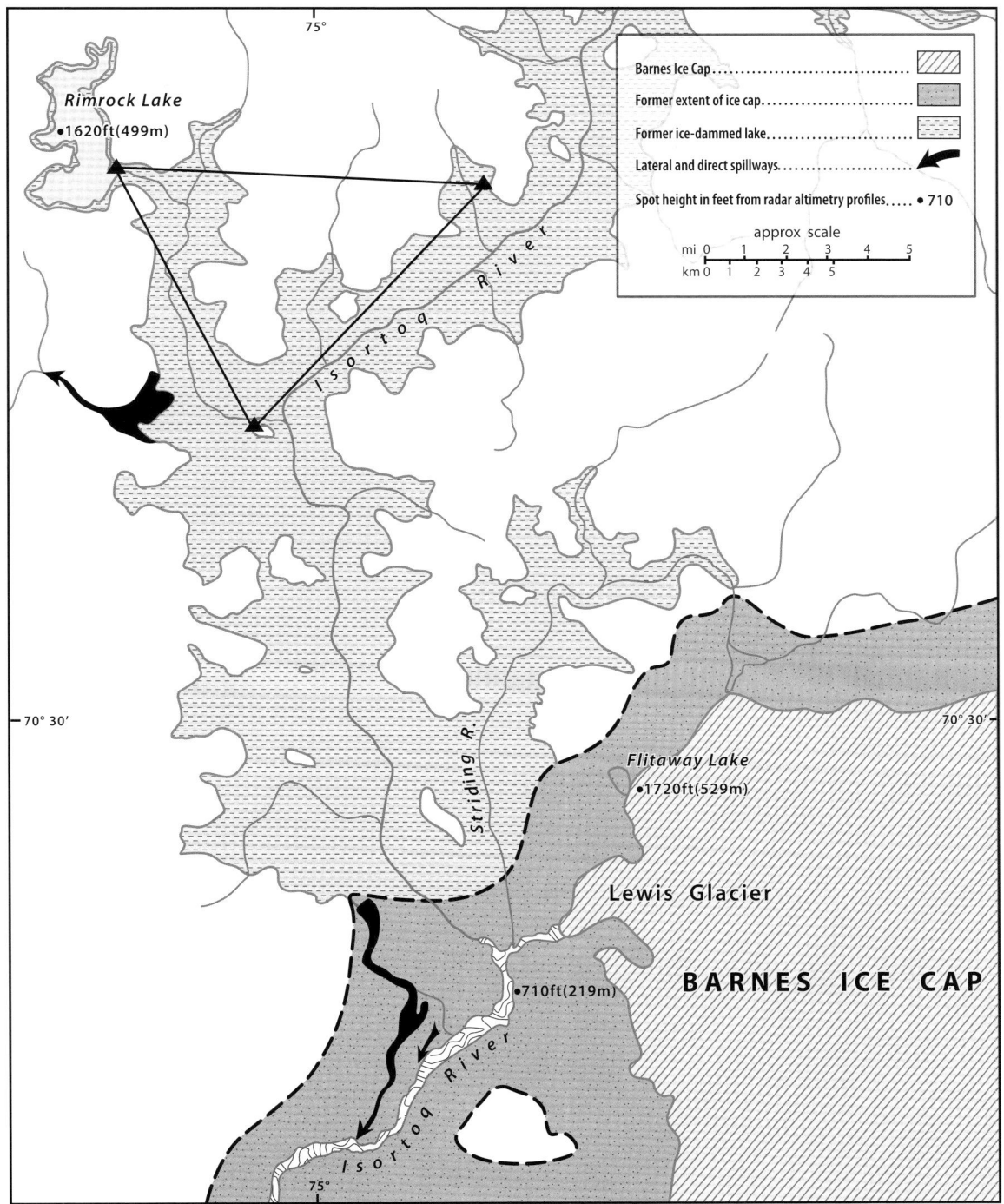

MAP 4: The most intensely studied area during the 1961 reconnaissance that extended from the northwest margin of the Barnes Ice Cap to Rimrock Lake. The map displays the conjectured area of former Glacial Lake Lewis, dammed by the Barnes Ice Cap when it extended to block the middle reaches of the Isortoq River. The approximately equilateral triangle illustrates the precise levelling work undertaken to determine whether the ancient lake shoreline had experienced isostatic tilting after the lake had drained with retreat of the ice cap. (Source: Modified from Ives, 1962, p. 199.)

Fig. 5: Glacial Lake Lewis shoreline. Wave and ice-push action, along with the outflow of water as the lake drained, has swept away all but the largest boulders from the surface below the shoreline. Beyond it the interminable boulder fields seem to extend forever. (Photo: July 1961)

Fig. 6: Rock lichens with a nickel for scale. The greenish-yellow lichens are of the *Rhizocarpon geographicum* group of very slow-growing and long-living lichens. (Photo: July 1963)

areas, and their distribution appeared to have been influenced by a prevailing northwest wind that had caused wind-drift of snow due to the topography. We could easily imagine great drifts of snow on the lee side of hills and ridges that had persisted through entire summers during some period in the past. Extensive areas of the interior plateau and the slopes of the Baffin Island coastal mountains had been blanketed by thin patches of permanent ice and snow comprising more than half of the total land area.

The next questions were these: At what time had the thin snow and ice cover been formed? And for how long had this period of near-glacierization persisted? The margins of lichen-free areas were studied closely. Numerous counts of the size of individual lichen diameters and their total number, as well as the percentage of lichen cover, were made at numerous sites on each side of this prominent shoreline. I hypothesized that north-central Baffin Island had been

Fig. 7: Glacial Lake Lewis drainage. With the collapse of the dam formed by an expanded Barnes Ice Cap about two to three hundred years ago, the massive outflow of water swept away much of the loose surface material. (Photo: August, 1961).

on the threshold of "instantaneous glacierization" some two to four hundred years ago (Ives, 1962). Consultation the following winter with Dr. Roland Beschel, the leading lichenologist, would help confirm this interpretation. (Fig. 6)

As walking conditions improved with the now rapid snowmelt, we were able to progress farther and farther afield from base camp. We established light overnight camps up to twenty kilometres to the east, and our small rubber dinghy enabled us to cross the Rimrock River below base camp and examine a long stretch of the north shore of the Isortoq River to our south. We traced other former lake shorelines with glacial lateral deposits at even higher levels, and a series of abandoned lake spillways was located through which these former lakes must have drained. The former drainage pattern led mainly westward, into the upper valley of the Freshney River[5] and thence into Foxe Basin. (Fig. 7) Concurrently, we began

a detailed study of the fabric of the loose material that comprised the multiple ridges below the Glacial Lake Lewis shoreline. The study was limited by the presence of frozen ground (permafrost) at quite shallow depths. The ridges have since become known as "cross-valley moraines" (Andrews, 1963). (Fig. 8)

To this point, we had at least partially explained the three sets of features that I had originally stumbled across while working on the *Arctic Pilot* during the 1956–1957 winter in Ottawa. The light- and dark-toned pattern had been produced by differential growth of rock lichens due to different time periods since their emergence from beneath permanent snow, ice, and lake cover. The cross-valley moraines were probably formed by a process of subglacial debris being squeezed up into the basal crevasses beneath the snout of glaciers flowing into ice-dammed lakes. While we had not been able to visit the major moraine system, which would become known as the

Fig. 8: Oblique aerial view of cross-valley moraines. The middle section of the Isortoq valley. (Photo: July 1962)

Cockburn Moraines, we realized that it would be of major importance for unravelling the history of retreat of the last great ice sheet. To attempt ground investigation this year would require a fully operational Cessna while there was still time for more distant fieldwork, before the onset of snowstorms and temperatures low enough to freeze the lakes for another long Baffin winter.

On the exigencies of Arctic bush plane charter

This was the first time in my career that I had had the resources to control use of a bush plane on floats. It was an exciting prospect considering the unexplored interior of Baffin Island and the impressive mountain and fiord landscape along its northeastern rim. However, the long delay and uncertainty of the missing Cessna's whereabouts was causing a high degree of frustration.

During a desultory attempt on August 4 at 8:00 a.m. to contact Fox-2 by radio, after the month-long silence, we were suddenly startled by a response. All three of us sitting on the edge of Rimrock Lake leapt off the ground in excitement as Vic Sim's voice came through loud and clear: "Jack, the Cessna has reached Frobisher and is awaiting improvement in the weather for clearance to proceed northward, but we are socked in." Overwhelming disappointment returned, and the waiting game resumed. A strong southeast wind with intermittent rain and snow had dispatched the remaining lake ice to the northern end of Rimrock Lake, leaving plenty of open water for the Cessna to splash down and so reach us, but a faint radio call in the early evening indicated that the Cessna was still at Frobisher.

The weather rapidly improved all through the day of August 5. We had excellent radio contact with Vic at noon and were delighted to receive an enthusiastic account of the progress of his own fieldwork. I suggested that he and Claude Lamothe visit us as soon as the Cessna was available. By suppertime our weather was clear and sunny. We had packed and were ready for our first planned airborne camp move—to Flitaway Lake[6] and the north end of the Barnes Ice Cap.

The morning of August 6 was also perfect for flying, and we had another clear radio contact with Vic. But his news was even more disheartening: apparently the earlier information that the Cessna was at Frobisher was false, and its whereabouts were unknown. Frustrated in our original plans, we spent the day hiking along the main lake shoreline, albeit adding a wealth of minute detail to our notebooks. The walk back to camp in a wind freshening to fifty to sixty kilometres per hour was arduous, and we had to proceed carefully over the unstable boulders so as to avoid broken or sprained ankles.

The first definite but still dismal news of the Cessna reached us on August 8. Vic had phoned Keith Fraser at the Geographical Branch in Ottawa and learned that the Cessna was in fact grounded at Wakeham Bay in northern Quebec. The connection with Keith had been very poor, so Vic was not sure whether the Cessna's delay was due to mechanical failure or poor weather. But wait! Another radio message was received on August 9 from Vic, reporting mechanical trouble at Wakeham Bay and telling us that a new propeller was being shipped from Ottawa to Fort Chimo. With this truly disappointing news, I asked Vic to enquire about the possible charter of a light aircraft from Wheeler Airlines at Frobisher Bay. However, fog and snow enshrouded us the following day, with visibility ranging from zero to about four hundred metres. Then at 5:00 p.m. came the unexpected news by radio that the Cessna had landed unannounced at Fox-2 mid-afternoon and should arrive at our Rimrock camp in about seventy-five minutes. Would this finally be the day?

We prepared once again for immediate camp transfer and, believe it or not, by 6:30 p.m. we were shaking hands with Vic, Claude, and our newly arrived pilot, Robbie Levesque—the first fresh faces we had seen in seven weeks. Robbie, the cheerful and very youthful pilot, was more than a little abashed over the course of events. He was trying hard to make up for our disappointment with enthusiasm, although it was apparent that he was not at ease with his surroundings. Nonetheless, he proposed to fly immediately with a full load of equipment to Flitaway Lake, some forty kilometres away, and then return for us on subsequent flights. I had a deep-seated premonition that I should insist on accompanying him.

We had requested a pilot also qualified as an aircraft mechanic or engineer as part of the original charter contract. Robbie had no training as a mechanic, and furthermore, no previous bush or Arctic flying experience. He had brought a mechanic with him, Dave McEwen. However, for Dave to fly along in the Cessna on each trip would mean a severely reduced Cessna payload. So on his own initiative, Robbie made arrangements for Dave to remain at Fox-2. The downside was that not only would this setup deprive us of his services for most of the time, but his accommodation fees charged by the DEW Line would make a good-sized dent in our budget.

More disturbing observations now surfaced. It was apparent that the Cessna's engine was leaking oil and the port pontoon was taking on an uncomfortable amount of water. The starter motor was functioning only intermittently—this meant that, on occasion, the engine could not be started by the pilot from the cockpit but instead with the assistance of some courageous soul standing on the narrow front tip of a pontoon and "swinging" the propeller while balancing just above very cold water. So the process began; after several swings of the prop until the engine eventually fired, I had to crawl along the pontoon with head down so as not to be decapitated by the now accelerating prop and climb into the co-pilot's seat. A further problem was that the Cessna's radio was not at all reliable, a feature that came to haunt us later on. This was certainly an unexpected and shaky beginning for my first aircraft charter, but much worse was in store.

At last we were gaining speed across Rimrock Lake and Robbie steadied the Cessna for takeoff. It was very fortunate that the lake extended about five kilometres north to south, as our engine seemed to be significantly underpowered. I seriously thought we would reach the far end—and an abrupt, rocky conclusion to life—before Robbie would manage to get the pontoons off the water. But he literally pulled it off, and there followed an unforgettable low-level flight across such an incredible array of glacial landforms that my earlier frustration was replaced instantly by exhilaration.

Now we were circling Flitaway Lake. (Fig. 9) Robbie thought it rather a small body of water, but there was no other. After quick discussion, a sudden aura of confidence pervaded the cockpit and we decided to put down. There would be no trouble in landing; the test would be on the takeoff.[7] After a perfect landing, Robbie shut down the engine and leapt out onto the pontoon, looped a rope around a strut, and readied to jump ashore. I assumed that he was pushing hard to make up for the lost time, but the arrival strategy was planned. The plane's forward motion would carry him close enough to the shoreline for a jump into the water, and with a couple of splashes forward he would be on dry land, holding the plane by the rope.

The lake was opaque—hardly a surprise to me, considering that it was an ice-dammed lake. I tensed when I realized what Robbie was proposing to do. I called out, "Be careful, Robbie! There may be buried ice and melt-holes." Robbie called back, "It's shallow . . . not to worry!" as he jumped off the pontoon. Then, with a great splash, he disappeared, completely submerged beneath the surface of the pea soup "shallows."

Obviously Robbie had never enrolled in Physical Geography 101. He was wearing waterproof waders that went up to his waist. In desperation, I leapt out of the cockpit and threw myself onto the pontoon. Straddling it, I managed to grab one of his shoulder straps as his head and shoulders popped out of the water. I then learned the weight of a man with waterlogged waders who needed to be pulled onto a pontoon rather quickly. The water was as cold as fresh water can ever be. By the time I had him draped over the pontoon, the breeze had pushed us back, well out from the shore. Robbie was turning blue but assuring me he could hold on while I tried to restart the engine. Mindful of our earlier experience with starting the Cessna, my heightened anxiety returned; there was no way I could swing the prop on the outside and open the throttle at the same time from inside the cabin, while Robbie, dripping wet and shaking, was in absolutely no condition to assist. Somehow, with much fumbling, I succeeded in starting the engine, this time directly from the pilot's seat. With a huge sigh of relief, I opened the throttle and the Cessna surged toward the shore, thrust its pontoons in the dry sand, and stalled, fortunately hard and fast. But it was not quite over for us. Without wasting any time I had to remove Robbie from his waders, pitch a tent, light a primus stove, and pull him with me into a sleeping bag. We had been on the edge of disaster in our first few hours of airborne geographic exploration, but had risen to the challenge.

Robbie warmed up quickly. I squeezed out of the sleeping bag and made us both mugs of hot instant coffee. He recovered rapidly, asked me not to mention his dunking to the others, and was quickly airborne and on his way back to Rimrock Lake to pick up the rest of the crew, together with more food and equipment—two fifteen-minute flights each way. And although he had to use every available metre of surface water, he took off without a hitch. Nevertheless, when we considered that we would shortly need to refuel, yet another problem surfaced. The original charter document had stipulated that the Cessna must bring with it from Ottawa two empty ten-gallon fuel drums for use in refuelling on remote lakes. When I brought this to Robbie's attention, his face fell and he had to confess that he had never been informed of that requirement. Now that he was with us and understood that we would have to do

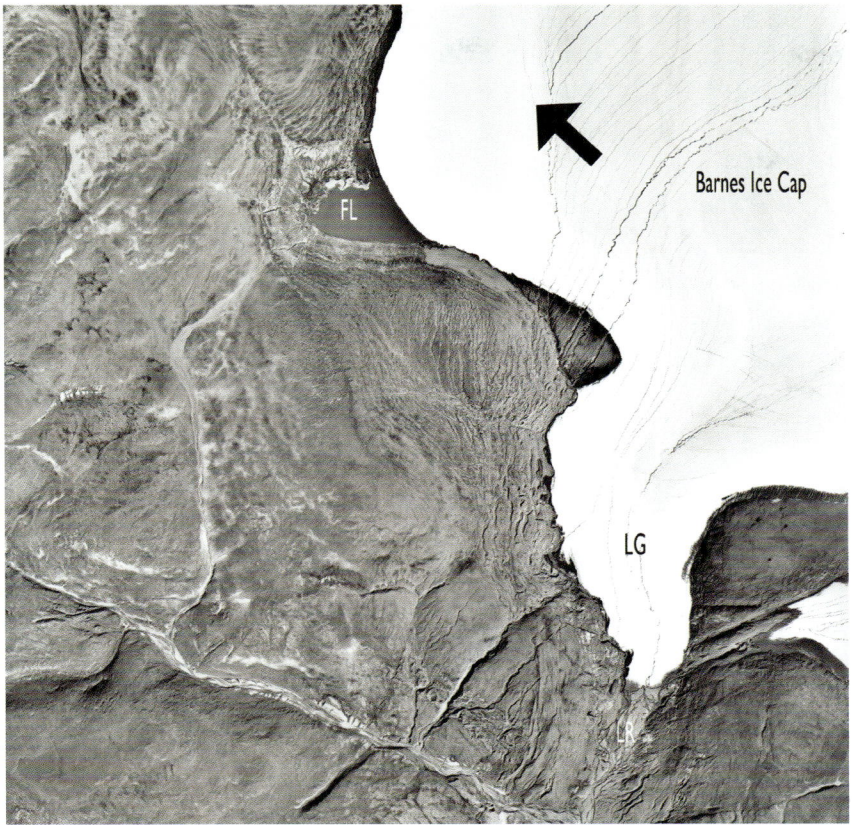

Fig. 9:
Aerial view of the northwest margin of the Barnes Ice Cap with its small outlet glacier, the Lewis Glacier (marked LG). Flitaway Lake (FL) drains along the edge of the ice cap and beneath the northwest side of the Lewis Glacier to emerge from the glacier snout as the Lewis River (LR). The numerous dry channels subparallel to the glacier margin and the higher shorelines around Flitaway Lake are indications of progressive glacier thinning and retreat. Arrow points north. (Source: Modified from Ives & Kirby, 1964, plate 1)

all our refuelling directly, and somewhat inefficiently, from the forty-five-gallon drums cached at Rimrock Lake, losing the flexibility of extended sorties using the smaller drums, he realized the problem. He also eventually had to explain to the others why he was soaking wet. I could hardly suppress my laughter. Despite the much too adventurous beginning of our aircraft charter, by late evening all six of us were well fed and tucked into sleeping bags by the peaceful margin of Flitaway Lake, with the ice cap grandly reflected on its calm surface.

Flitaway Lake: To flit or not to flit?

During preparations in Ottawa the previous winter, I had realized that the place we came to name Flitaway Lake would become a focal point in subsequent operations, should the 1961 reconnaissance justify the long series of expeditions to north-central Baffin Island that I hoped to organize. We were, therefore, quite concerned about the lake's stability. The abandoned high shorelines that surrounded the lake indicated that its level had been falling for some time, and the overflow channels down the side of the Lewis Glacier, apparent on the air photos, gave a good indication of where the water was going. As a potential site for the projected 1962 base camp, its rate of lowering was critical. One of our immediate objectives, therefore, was to determine whether or not we had made a safe choice.

After breakfast the next morning, our first priority was a return to Rimrock Lake for refuelling. Leaving the others to reconnoitre the surrounding area, especially to find access onto the Lewis Glacier and to test the ice drill, Vic and I took off with Robbie. We planned an extensive reconnaissance as far west as the lower Rowley River and the coast of Steensby

Inlet, Foxe Basin. As we were buckling our seat belts for takeoff, Robbie explained that he had paced the length of the lake before breakfast and found it was slightly less than minimum specifications for takeoff of a Cessna on pontoons. The specs, I thought, were obviously intended to apply to a Cessna that would meet general flight regulations. Yet it was apparent, from our experience so far, that our Cessna was conspicuously lacking in performance. So it was very fortunate that Flitaway Lake had gentle approaches with no steep slopes close by and that wind conditions remained favourable when needed—a fairly strong breeze along the length of the lake, into which the plane could take off and land.

We lunched at Rimrock Lake after refuelling. As the engine once again did not respond to Robbie's efforts from the cockpit, Vic did the honours with the propeller; at least he could swim, although gamefully he explained that he was probably just as subject to cardiac arrest as I was. We failed to make radio contact with Fox-2 to register our flight plan and so had to depend on the preliminary plan we had filed during our earlier 8:00 a.m. sked from our base radio, now at Flitaway.

From Rimrock Lake, Robbie flew south along the course of the Freshney River to its confluence with the Isortoq River and followed it as far as the lower end of Isortoq Lake before turning north toward the lower Rowley River. Along the way we located extensive moraine systems in the lower Isortoq valley that we thought were associated with what appeared to be an extensive marine limit (that is, a formerly higher sea level) where Isortoq Fiord entered Grant-Suttie Bay. Vic and I agreed that he and Claude should attempt to examine the area on the ground after John, Peter, and I were established close to Steensby Inlet on the lower Rowley River. This major river emerges from an impressive gorge, and its lower course runs through a series of lakes before finally entering Steensby Inlet. My immediate objective was to locate a suitable campsite for the next phase of our summer's fieldwork.

We found a perfect sand-girt embayment on the upper part of a lake that we later named Separation Lake.[8] It was ideal for floatplane operations and looked like a good campsite, so Robbie put down and we had a relaxed second lunch by the plane. In deteriorating weather, we then took off and followed the Rowley River down to the coast; from there we flew to the lower end of Isortoq Lake and up the Isortoq River back to Flitaway. We landed without incident.

During our absence, John had discovered that an essential piece of the ice drill must have disappeared in transit from Ottawa. Vic promised to test the resources of the Fox-2 machine shop to see if a new handle could be quickly improvised. Then Robbie, with Vic and Claude, took off in the Cessna for the amenities of the DEW Line while John, Peter, and I—the original "Rimrock Three"—enjoyed a quiet evening watching the cloud base gradually drop to lake level. Light snow completed the ethereal sense of Arctic gloom as we readied for sleep.

The 8:00 a.m. radio sked with Fox-2 on August 12 revealed that the Cessna had not arrived there from Flitaway the previous evening and that its current whereabouts were unknown. Then followed the ominous request from Lou for details of the flight: time of departure; names of passengers; estimated fuel range; and our own latitude and longitude. We arranged a further radio sked for noon. Lou, the station chief, would determine whether and when to call for a search and rescue mission, though I did alert him that, as the Cessna's radio was likely inoperable, the lack of contact with the plane might not be as serious as it seemed.

Regardless of the nerve-wracking uncertainty, I decided to do a day's fieldwork with John, leaving Peter to monitor the radio. We took a long, brisk walk down the side of the glacier and along the meltwater outflow (Lewis River) to its confluence with the Striding River. The weather all around us was threatening—squalls, heavy cumulus clouds, scattered rain. We were blessed by a local enclave of high pressure generated by the ice cap and spent most of our day in sunshine. Our route took us through an abundance of glacial features that emphasized the good choice of field area.

Halfway back to camp, we were relieved to see our Cessna pass low over us, heading in the direction of Flitaway Lake with no apparent problem. We reached camp at 7:15 p.m. and learned the details of yet another escapade. Peter told us that the previous evening Robbie had been forced by fog to land on a lake barely thirty kilometres short of Fox-2. Later, in a partial clearing of the fog, they had made it a further fifteen kilometres—but this proved to be a mistake. While the first forced landing had brought them down on water with a reasonable lakeshore for tethering overnight, on the second lake they had been obliged to spend a very uncomfortable night in the plane, tied to a large rock in the middle of the lake and praying for continued calm conditions. They had reached Fox-2 late the next morning, just in time to alert Peter on the noon radio sked that they were all safe.

Lou Riccaboni and another DEW Line staff member had accompanied Robbie to Flitaway "for the ride." They had brought with them a very quickly improvised new handle for our ice drill, as well as the first mail from Ottawa that we had received since our departure on June 14 and some much appreciated fresh bread with real butter. Once again, the Cessna's radio had failed, so Robbie had been unable to inform Fox-2 of his position the evening before. Thus passed another day of unwelcome excitement. We settled down to a night of light snow with the temperature barely above freezing.

We continued to familiarize ourselves with the immediate surroundings of Flitaway Lake and the northwestern margins of the Barnes Ice Cap until the evening of August 15. The glaciological potential of the immediate area was assessed; four holes about 150 centimetres deep were drilled, and stakes for ice movement measurements were inserted for resurvey the following year. Cairns were built at carefully measured distances from the northern ice cap margin for future determination of change in conditions.[9]

That evening, Robbie arrived on schedule, landing the Cessna at seven o'clock and taxiing to our camp with what was now well-practised skill. I climbed on board with a tent and minimum equipment and food. We flew directly along the Lewis Glacier, followed the Isortoq River to Isortoq Lake, and then swung across country toward Separation Lake, landing smoothly like experienced veterans. A quick turnaround and Robbie was airborne once more by 8:30 p.m., returning to Flitaway to fetch John and Peter to join me. I pitched the Arctic Guinea tent and prepared a site for the Mount Logan tent in anticipation of the Cessna's pending arrival. Then I sat down to wait patiently, nonetheless watching with apprehension as clouds formed on the surrounding hilltops.

Around 11:00 p.m. on August 15, I turned in, tired of waiting and hopeful that this would not turn out to be another minor (or major) catastrophe. I must have fallen asleep instantly—my youthful response to tension! It seemed almost immediately that a plane was approaching. A few minutes later, the Cessna landed, through deepening twilight but fortunately with enough of a breeze to cause ripples on the water, so Robbie could gauge how far he was above the lake's surface as he came in to land. (Pilots hate totally calm lake water, which can bring on a special kind of vertigo and a potentially serious failure of depth perception.) A hearty welcome, a quick brew of coffee while the Logan was being pitched, and we were all in our sleeping bags by 2:00 a.m. As "summer" was nearly over, we were beginning to experience a few hours of darkness.

Over breakfast, we discussed the previous evening's near disaster. Robbie had picked up John and Peter plus all the necessary equipment from Flitaway. He had then flown to Rimrock to refuel. A stiff northerly wind made refuelling rather difficult, as it kept pounding the Cessna on the rocks, causing the need for constant pushing off. When the Cessna's

Fig. 10: Raised marine beaches along the coast of Foxe Basin. (Photo: August 1961)

tanks were full and all were ready for takeoff, the engine refused to start—a dead battery, or the generator refusing to cooperate. After a lengthy struggle, Robbie managed to connect directly to a spare battery that he had wisely picked up from Fox-2 and that saved the day.

We continued to discuss our Cessna problems until it was time for Robbie to make his way back to Fox-2 to assist Vic. We all recognized that the Cessna was likely being flown in defiance of a number of government regulations. The real danger was that it would make an unscheduled landing and then be unable to start. Without a functioning radio, it would be stranded in a location not known to the DEW Line personnel. To bring this danger home even more forcibly, we had another struggle with the starter motor and had to resort to the spare battery again.

Robbie was airborne by eleven o'clock that morning, and John, Peter, and I set out to reconnoitre the new territory. We had walked scarcely three kilometres when we spotted the Cessna heading back to our camp, so we hastily returned. This time it was due not to an aircraft malfunction but to the thick fog shrouding Fox-2 and preventing the use of our usual lake landing spot. Yet another day of fieldwork was at risk of being wasted. In such situations, flexibility in field operation can salvage precious time; that day, I decided to take advantage of the Cessna's presence to make an aerial reconnaissance of the Steensby Inlet coast and the downstream section of the Rowley River. John and Peter, meanwhile, set off on foot to take a look at the hill country to the northeast of the lake.

It took ninety minutes to start the plane, reminding me once again of the precariousness of our situation. Nevertheless, the low-level aerial views along the Steensby Inlet coastline were both impressive and helpful. In very good light, I took many photographs of the large moraine system to the southwest of the camp and of the beautifully formed flights of raised marine beaches along the Steensby Inlet coast. (Fig. 10) Not surprisingly, though, Robbie was reluctant to land for fear of being unable to restart the Cessna's engine—and of having no juice to power his radio, either.

Robbie returned me to camp and went on his way to support Vic south of the Barnes Ice Cap for the next several days. This gave us four days of peaceful isolation that facilitated long walks in all directions. Apart from garbled radio mutterings in the evenings, we were completely and thankfully cut off from the outside world for a while. Yet, as we were soon to discover, our logistical support was about to reach another dead end.

Further airborne drama

It was a very poor radio contact late on August 20 that jerked us back from that peaceful solitude into the real world of aircraft operations in the Far North. While initially the news was not at all clear, we gathered that the Cessna had finally become inoperable, leaving Vic and Claude marooned somewhere between the Barnes Ice Cap and Fox-2.

As there was nothing we could do to relieve the drama developing to our south, we managed another fine day of fieldwork. However, as we headed back to camp, we saw a helicopter heading in the same direction. We arrived to find Vic and Harold, the helicopter pilot, awaiting us. The helicopter was on charter to a gravity survey team of the Dominion Observatory working south of the Barnes Ice Cap. From Vic we quickly learned of the sad necessity of its appearance so far off course.

Here's what had happened. The previous day, Robbie had transferred Vic and Claude from Generator Lake to Flint Lake, had returned to Fox-2 to refuel, and was then set to head back to Flint Lake, pick the men up, and take them north along the Foxe Basin coast. Unfortunately, the Cessna chose that moment to enter another of its infamous sulks. It was reluctant to lift off the lake's surface; indeed, a dangerous crash into the barrier of large boulders that fringed the lake appeared unavoidable. At the last minute, Robbie somehow managed to leapfrog the line of boulders and plop down on the far side, landing heavily in a tundra marsh. This last ditch attempt to save our air contingent had one snag—a large boulder in the marsh promptly tore off one pontoon and severely holed the other. Robbie, fortunately, was not injured, but Vic and Claude were on Flint Lake with neither food nor tent, and the rest of us in the great northern unknown were not yet even aware of the drama.

Fox-2 station chief Lou, meanwhile, had alerted our Dominion Observatory friends conducting the gravity survey south of the Barnes Ice Cap—hence the helicopter "rescue." In the meantime, Robbie had telegrammed his HQ at Bradley Air Services in Ottawa:

> Extensive damage to floats, Cessna HLR near Longstaff Bluff, NWT during aborted take-off. Nil injuries. Require replacement aircraft on floats to complete contract. Full report to follow. Please advise your decision soonest.

Vic indicated that it would take seven to ten days for a replacement aircraft to reach Fox-2. Thus a change in plans was essential. We quickly worked out a flexible scheme that would cover "all eventualities"—assuming such is ever possible. The helicopter would fly John and me to a lake close to the Steensby Inlet coast with all the scant food we could muster, including two bars of chocolate generously offered by Harold, the helicopter pilot. Harold would then return south with Vic and Peter. By separating from Peter, our slim rations would last somewhat longer.

Next, Vic and Peter would join Claude, and from there they would work up the Foxe Basin coast toward Grant-Suttie Bay by canoe and foot. John and I would continue our fieldwork while awaiting the new plane from Ottawa. If an emergency arose, the Dominion Observatory team would send their chartered Norseman floatplane to pull us out.

These arrangements, though hastily assembled, left me with John in a comfortable camp within easy walking distance of the coast and the distinctive flights of uplifted marine beaches that were one of our major objectives of the summer. As it turned out, we were granted a week of excellent conditions for fieldwork and were able to dismiss all thoughts of the Cessna and radio disruptions for a while. I was confident that we would complete the vital parts of our planned fieldwork and that we would be "rescued" one way or another. I did not have much faith that we would ever see a replacement Cessna, in light of this summer's experience, but if we could squeeze in three or four days of floatplane support, we would be able to inspect the Cockburn Moraines and the fiord heads far to the east.

But all was not well. Unknown to us, the final disaster of the air odyssey was unfolding far to the south. It was only when we made our ultimate return to Fox-2 that we learned from Robbie's telegram that a Cessna had been promptly dispatched from Ottawa by the good Mr. Bradley of Bradley Air Services. However, it had been dispatched to the North with a rather serious flaw, to say the least. Having no replacement pontoon-equipped Cessna, the company had transferred the pontoons from a Piper Cub to a Cessna that, until then, had been equipped only with wheels. Even I would have assumed that a Piper Cub was appreciably smaller than a Cessna and that, consequently, its pontoons would likely be smaller. This proved to be true, and it spelled trouble.

The Cessna was flown along the Ottawa River to Montreal, then along the St. Lawrence via Quebec City to Sept-Îles. From there, it followed the railway north to Schefferville and onward to Fort Chimo,

where the last refuelling was completed. What followed should have been an easy tourist-style flight up the west coast of Ungava Bay to the settlement of Payne Bay. Now for the excitement. The pilot made a perfect landing on a calm sea and taxied in toward the jetty. While negotiating the final turn, to come alongside its anticipated mooring, the centrifugal force of the turn unbalanced the aircraft on its undersized pontoons. It capsized—and sank. Fortunately, the pilot could swim. When news of this latest and final mishap reached station chief Lou, the Dominion Observatory geophysicists were called in. An hour or so after receiving the call, Greg Lamb landed his Norseman on what we had come to call Windless Lake and taxied in to our camp to give us the amazing news. While this tale of woe and incompetence was unfolding far to the south, John and I, totally oblivious to it all, had enjoyed a week of carefree fieldwork.

Final days of Baffin '61: Fieldwork at Windless Lake

The Windless Lake camp proved a perfect location for what became the final reconnaissance work of 1961. We were encamped about two kilometres from the sea, at a site accessible by a short pass through the coastal hills. The hill crests dropped quite steeply, from heights of up to two hundred metres to a narrow coastal fringe. A virtually unbroken line of subparallel raised beaches bordered the coast as far as we could see in both directions. We quickly set to work with two surveying altimeters to determine the height of the upper marine limit, which was very distinct in this area. There was an almost straight-line division between hummocky glacial moraine above and sand and gravel beaches below. We were able to extract from the raised beaches small samples of marine mollusc shells for possible radiocarbon dating. This was very time consuming because of the scarcity and small size of the shell fragments, but it was

Fig. 11: Inuit spearing char in the estuary of a small stream draining into Steensby Inlet, northeastern Foxe Basin. (Photo: August 1961)

worthwhile—we later obtained a time sequence of uplift of the land and an estimate of the date when the postglacial sea entered this northeast corner of Foxe Basin.

We extended the reconnaissance onto the delta of the Rowley River and inland to include a complete circuit of Windless Lake. We also walked down the outlet stream of the lake southwestward to the sea, where we met the only local people we had seen during our entire summer on Baffin Island. Two families of Inuit had anchored their umiaks in the small estuary. The older males were spearing Arctic char in the almost still waters of the stream, and we were very impressed by their skill. (Fig. 11) They rarely

missed a target, and a pile of char accumulated on the shore. A youth in a small boat was manipulating a net and adding to the catch. The women and several children either remained afloat or milled around on the edge of the stream. They seemed quite shocked by our sudden appearance but were very curious and friendly. Even collectively they were able to muster only a few words of English, although we understood that they were on their annual autumn fishing and hunting trip from Igloolik, on the far side of Foxe Basin. The only incident was when John—presumably because of his dark, hairy face—frightened one of the small boys into a panic of screaming, an event that added to the laughter and conviviality all round. The umiaks were already well loaded with seal, fish, hare, and foxes. One of the men offered me a spear that I eventually took home with me as a gift for Tony, my young son who was not yet six months old. As a parting gesture, they invited us into one of the umiaks so that we were able to cross the stream dry-shod. After a hearty farewell, we continued to walk northward along the coast and eventually back to camp.

Our levelling of the marine shorelines recorded a maximum height of ninety-six metres asl[10] and we collected two samples of seashells from heights of fifty-five metres and twenty-two metres, respectively. To these we were able to add three other samples collected by Vic farther south, enough to demonstrate the progressive uplift of the coast over the last 6,700 years. The field data and the radiocarbon ages would also indicate when the sea had invaded this part of the coast following the eastward withdrawal of the remnant continental ice sheet.

With an abundance of good weather and excellent walking conditions, compared with the torturous boulder fields of the Rimrock Lake area, time passed very quickly. Our only local companions were Arctic hare, all white and startlingly conspicuous against the darker terrain.[11] On the evening of August 28, preparations for our evening meal were interrupted by the shattering noise of an aircraft that was obviously intent on landing on Windless Lake. And so the Lamb Airways Norseman piloted by Greg Lamb arrived. It carried as its passenger Angus Hamilton, leader of the Dominion Observatory's geophysical party.

We discussed our prospects with Angus. As it seemed unlikely that Bradley Air Services would ever reach us with a replacement Cessna, and considering our distance from his base camp, Angus recommended that we be evacuated to Dewar Lakes (Fox-3). If a replacement Cessna were to reach us, we could easily fly from Dewar Lakes to examine sections of the Cockburn Moraines. This proposal by Angus was eminently reasonable, and considering the lateness of the season, I gratefully agreed.

We hastily broke up the camp, loaded everything into the Norseman, and took off with a great roar into the gathering dusk. We flew along the entire length of the Barnes Ice Cap and landed at Dewar Lakes in the moonlight. The impressive base camp, by the lake, sat below the hilltop on which stood the Fox-3 DEW Line station, about three kilometres distant.

In comparison to the lightweight dried food we had been eating for nine weeks, breakfast the next morning was a delicious treat: hot fresh bread with butter, toast, fried eggs, bacon, sausages, oranges, real coffee, and above all, convivial fresh faces and an enthusiastic sharing of experiences. The Dominion Observatory standard of field living was certainly higher than ours—although we prided ourselves on needing far less time for meals.[12] After breakfast, Angus offered me use of the Norseman in the tidying up and evacuation of our Rimrock Lake base camp. I gratefully accepted and flew the two hundred kilometres northward with Greg Lamb and Roy, his engineer. We quickly loaded all the valuables, cached the surplus food and fuel for another year, and readied to take off.

We headed directly for the Barnes Ice Cap. Greg had kindly offered to fly me along its entire 145 kilometres of crestline, this time in full light. This gave me an excellent overview of the entire ice cap, valuable for future research plans. We landed at the geophysicists' base camp in calm conditions and settled down to another evening with our generous hosts.

The next two days brought low temperatures and intermittent snow. The ground began to turn white. I was able to talk by phone with Robbie at Fox-2; there was still no news of a replacement Cessna. Already the lakes were beginning to freeze—even the large Dewar Lakes had ice around their margins. It was mid-afternoon on September 2 before reasonable weather at both Fox-2 and Fox-3 allowed Greg to undertake the forty-minute flight to get us away. As we passed low over the rolling hills we could see that many of the smaller lakes had already frozen, and the land was white. Even the lake below the Fox-2 station

Fig. 12: Standing (from left): John Andrews, Jack Ives, Claude Lamothe, Vic Sim. Sitting: Peter Hill. (Photo: September 1961)

that we had used throughout the open-water season was frozen, forcing us to land on salt water in Piling Bay close under the radar station.

Unfortunately, we had landed shortly after high tide and had to struggle in very cold water to unload while keeping the Norseman afloat. We thought Greg would be able to take off if we pushed the Norseman farther out each time its pontoons began to touch the rocks below. But he also needed to refuel, and the rate of tidal lowering eventually caught the plane and firmly grounded it. That left Greg and Roy with an eight-hour wait for a return of the sea, which would restore their takeoff space.

As we were all struggling unsuccessfully to refuel the Norseman, Lou Riccaboni drove a truck down to the coast to personally take us and our equipment back up to the warm accommodation of Fox-2. We were to share a hut with Robbie and Dave McEwen, the mechanic, whom we were meeting for the first time. It was Lou who told us that, a couple of hours earlier, Robbie had learned by telegram the full details of how our replacement Cessna had met its watery demise. That removed any further question of continuing with the fieldwork. All that remained of the 1961 "summer" was to join up with Vic, Claude, and Peter and for us all to head south to warmer climes.[13]

The next day, Vic, Claude, and Peter walked in—very tired but contented, despite having to cache their canoe and heavier equipment and to walk the final fourteen kilometres from the base of Baird Peninsula. From that point onward, our travel was straightforward: a DEW Line lateral flight to Fox-Main, and commercial flights from Hall Beach airport via Montreal to Ottawa. It was 2:00 a.m. on September 7 when I knocked on the front door of our home in Ottawa to awaken Pauline to let me in. Peter and John had stayed in Montreal. Peter preferred sleep to another plane flight. And John had a special interest in meeting Martha, a McGill graduate student whom he was to marry two years later. The only real dent in my enthusiasm for the summer's experience came from my concern for our Cessna pilot, Robbie Levesque. He must have had very bad feelings about the series of misadventures, which had occurred through no fault of his own. He remained a cheerful and sturdy colleague throughout and performed admirably under very trying circumstances.

Evaluating aircraft operations

The first requirement on our return south after three months in north-central Baffin Island was relaxation with families and friends. As assistant director of the branch, I faced a backlog of work—one of the penalties of aspiring to lead field expeditions while undertaking the duties of a relatively senior administrative position. Dr. Nicholson proved very supportive, however. Dr. van Steenburgh, Director-General of Scientific Services, Mines, and Technical Surveys, requested a personal account. He seemed pleased with both the results I was able to relate and the way the rather challenging airborne support problems had been handled. He showed great interest and confirmed that I should develop long-term plans and keep him informed.

As for the airborne support, its successes and its problems and near disasters, it was agreed that the operation had been a valuable learning experience. Aircraft charters by government agencies were subject to an open bidding system, and Bradley Air Services, the winning contractor, had tendered the lowest bid. I was assured that next year's selection of air support services would be much more tightly controlled and that I would be personally involved in the process.

Vic and Peter reported for duty after short leaves, and both were a great help in sorting out and storing the equipment we had brought back. Vic prepared a detailed field report. John remained in Montreal, while Claude returned to his university studies. So ended the 1961 Baffin reconnaissance. In retrospect, it was remarkable that we had achieved nearly all of our original objectives, certainly enough to go forward for what proved to be the next six years.

CHAPTER 3

BUILDING THE TEAM
AND DEVELOPING CREDIBILITY

The 1961 reconnaissance had provided a mass of geographical and glaciological data, opening up the chance for careful planning on a much enlarged scale. It was apparent, however, that the field plan would have to be integrated with the reorganization of the Geographical Branch and the improvement of its long-term financial base. Dr. van Steenburgh's continued support would also be essential. A promising indication was that he had asked me for a detailed plan of operations to be delivered to him personally as soon as possible.

The breakdown of my activities for the next months, therefore, was clear: a detailed scientific agenda for the proposed fieldwork on Baffin Island and its presentation to Dr. van Steenburgh; enlargement of the permanent branch staff; and plans to justify this and to recruit experienced personnel capable of meeting the scientific objectives. In addition, there were a number of what became delicate, even controversial, administrative negotiations. Funding a multi-year operation in a distant and logistically challenging environment would have to be balanced with strengthening the other divisions of the branch, especially its Division of Economic Geography.[1]

The first major task was to establish within the branch structure an effective Division of Physical Geography, which would need additional staff committed to arduous fieldwork along with a well-equipped geomorphological laboratory in Ottawa. Frank Cook, a permanent member of the branch staff who had been a graduate student with me at McGill, offered to take temporary care of the development of the lab, for which we recruited a promising young man, Rolf Kihl, as technician. Frank had worked for several years on patterned ground and frost heaving in the Arctic; he would train Rolf and ensure acquisition of the equipment necessary for analyzing various types of soils.

Following the uncertainties, mental stress, and long periods of isolation of the summer, Peter Hill was quick to declare that fieldwork in the Arctic was not for him; he resigned, embarking instead on a career as a high school geography teacher. I was sorry for this but could hardly deny that he was making a wise choice. And so a vacant position became available. John Andrews was an ideal candidate, although civil service regulations of the time gave priority to Canadian citizens and John was a recent immigrant from the U.K.

The recruitment of John to a permanent position proved yet another learning experience; it would help me negotiate recruitment of several other specialists over the next few years. Since the process was vital to the success of the Baffin Island plans, it is explained here in some detail. In the 1960s, geographers had a lowly professional standing within the civil service, and the normally stringent rules of government recruitment were difficult to accommodate. For one thing, vacant positions were open to Canadian citizens only, although where a certain specialization was not available in Canada, non-citizens could be recruited from other countries. Veterans were given preference—and especially wounded veterans—provided their qualifications were similar to those of other applicants. In most cases, of course, such treatment was both ethical and laudable. Although it was incomprehensible that a wounded veteran would ever be expected to carry a heavy pack across glaciers in the Arctic, inappropriate appointments were a distinct possibility due to the perceived low status of geographers.[2]

John Andrews was not a Canadian citizen; rather, he had entered Canada as a graduate student less than three years earlier. Next, because academic rank was taken into account, a candidate with a doctorate would out-compete one with a master's degree (when all else was equal). John had just obtained his MA from McGill University.

Documents submitted by applicants were reviewed, candidates interviewed, and final decisions made by a committee composed of two senior officers from the recruiting department, representatives of the Civil Service Commission and the Department of Veterans Affairs, and one non–civil servant, usually a professional in the same or similar discipline—for instance, in our case, a geologist or historian would be suitable if no geographer were available.

The position left vacant by Peter Hill was advertised; I had drafted the job description, for a geomorphologist to undertake work in the Arctic that would involve long field seasons. Mary McCracken, our department's representative on the commission, was most helpful and I took her advice to restrict the competition to applicants living in the local region (i.e., Montreal-Ottawa-Toronto). For the non–civil service member, I was able to persuade Professor Bogdan Zaborski, chair of the University of Ottawa's geography department, to collaborate. Bogdan was an expatriate Pole who had been on the faculty of McGill during my doctoral studies and had served on my thesis committee, so I knew him as a most congenial colleague. Volumes could be written about Bogdan,

including his internment in the Siberian Gulag during the war. When released in 1946, he chose to head east rather than back west to Warsaw because he felt it was a geographer's sacred mission to learn as much as possible about "unknown" territory whenever he had a chance. So, after many adventures, he came to McGill via Vladivostok, accumulating hundreds of maps en route. Without him, John Andrews likely would not have entered "Her Majesty's Civil Service."

There were only a handful of applicants, and none were veterans, probably because few hopefuls aspired to "brutal" labours for long periods in the so-called wastes of the Far North. John had only a single serious competitor—another immigrant, but one with Canadian citizenship and a doctorate in economic geography. Several members of the adjudicating committee ranked him first and John second, on the basis of the written applications. Although I was concerned that the "top" candidate's expertise had no relevance to Arctic research, the fact remained that John had two counts against him: he was not a citizen, and he had only a master's degree.

As it transpired, during the interviews it became clear to Bogdan and the other members that the first-placed candidate was totally out of his depth when questioned about Arctic matters, geological, geomorphological, or even economic. John was offered the position and subsequently made a major contribution to the glacial history of Baffin Island, eventually becoming a world leader in his field. The lesson learned from John's recruitment proved very useful. It also seemed in retrospect that the process had been as much a test of me, because all subsequent recruitment of non-citizen specialists proceeded with relative ease.

Recruiting for leadership

The next step in my recruitment drive focused on Scandinavia and the coincidence of an invitation to lecture at Stockholm University. Shortly after returning to Ottawa from Baffin Island in September 1961, I was invited by Professor Gunnar Hoppe to visit Stockholm and to give a lecture on the results of the 1950s fieldwork in Labrador-Ungava and the 1961 Baffin Island reconnaissance. I was eager to do this as Hoppe's own research in the Swedish Arctic and Subarctic had provided me with many new ideas, and he had mentioned that many of the Scandinavian specialists in glacial geomorphology would attend. I arrived in Stockholm in early January 1962 and was a personal guest of Dr. Valter Schytt, whom I had met with Professor Hoppe during my participation in the now famous Abisko Symposium of July 1960.[3] In addition to the lecture, which prompted a lot of useful discussion, I had the opportunity to discuss plans for the Geographical Branch's Division of Physical Geography.[4] Dr. Gunnar Østrem, with whom I had also become acquainted during the Abisko Symposium, had developed an intriguing research approach to dating ice-cored moraines. He was immediately captivated by my descriptions of Baffin Island. He also spoke and wrote well in English, so I was delighted to discover that he was greatly attracted to the possibility of testing his new methods on the moraines of the Barnes Ice Cap.[5] This led to extensive discussion following dinner with his family, and I quickly realized I had a highly qualified new recruit who would greatly reinforce plans for a glaciology section in the branch. Even then, my subsequent formal recommendation for such a unit met with strong opposition from the GSC; specifically, many federal geologists seemed convinced that glaciology belonged firmly to the GSC.

As the departmental debate about glaciology expanded following my return from Stockholm, I was invited by GSC Director Dr. James Harrison to attend an annual meeting of the National Advisory Committee for Geological Research. Although I was given a very fair hearing, I was deluged with objections until an interjection from Dr. J. Tuzo Wilson, a pre-eminent Canadian geophysicist and a senior, non-GSC member of the national committee. He pointed out that, as the GSC had had more than a century to set up a glaciology section but had failed to do so, I should be given the opportunity.

The final decision rested with Dr. van Steenburgh. Dr. Harrison and I were to present our arguments during a private meeting with him. After a good deal of amiable, yet serious, discussion—which came down to Dr. Harrison's insistence that geography was not really a coherent discipline and, certainly, glaciology was an integral part of geology—I began to despair and to realize that I was fighting far above my weight class. As a last ditch effort, I quoted Tuzo Wilson's remarks to the effect that the GSC had done nothing about glaciology for the entire century of its existence and that "Jack should be given a chance." Dr. van Steenburgh exploded with laughter. He turned to Jim Harrison and indicated that he had decided in my favour.[6]

Gunnar Østrem was an excellent choice for head of the new Glaciology Section. First, however, I had to ensure his recruitment through the corridors of the Civil Service Commission. Further assistance from Mary McCracken led to a very specific technical job description plus a stipulation for fluency in a Scandinavian language.[7] The language stipulation required very careful justification as it was obvious that it would seriously limit any competition. Nevertheless, the formal process proceeded with amazing speed and Gunnar arrived in Ottawa the following July. Shortly after his arrival, he would leave with me for Flitaway Lake and the Barnes Ice Cap (see chapter 4).

Gunnar's appointment set in motion a number of related developments. Dr. Nicholson supported the enlargement of the Division of Physical Geography and its subdivision into three sections: Glaciology, Climatology, and Geomorphology. This was quickly approved by Dr. van Steenburgh. The bureaucratic channels still had to be negotiated, however, as well as support obtained for additional staff positions. Nevertheless, the process was well underway.

For the fast-approaching Baffin 1962 field season, I obtained approval for a second position in addition to Gunnar Østrem's. I was fortunate to recruit Brian Sagar. Brian was a geographer who had worked as a glaciologist in the Canadian High Arctic with Dr. Geoff Hattersley-Smith on the Gilman Glacier and on the northern Ellesmere Island ice shelf as part of the Defence Research Board's "Operation Hazen." Brian was enthusiastic to commit himself to a summer on the crest of the Barnes Ice Cap. We recruited two university students, Uwe Embacher[8] and Chris Bridge, as his summer assistants, early enough to arrive on the ice cap by May 18.

Undergraduate students as field assistants in the Arctic

As plans for the 1962 Baffin Island expedition began to fall into place, it became apparent that I would have to select and recruit very fit, intelligent, and willing young men; the women would have to wait another three years because of adamant departmental opposition. The McGill Lab was the ideal place from which to recruit graduate students. These students were especially well suited because the fieldwork most of them would have completed for their master's degrees were extensions of the research program I had established while director of the lab. Further, I had been an informal advisor to several of them, assisting with their selection of field research areas, and all of them had served as meteorological observers through a Subarctic winter—a decided advantage.

Recruitment of undergrads for summer employment was strictly formalized, especially as a summer federal appointment was a significant financial and possible career asset for the successful applicants. I was also conscious of the opportunity to use the Baffin field experience of a considerable number of

undergraduates and graduates to create a pool of strong candidates for future permanent Geographical Branch recruitment, as well as for strengthening the growing number of university departments of geography across Canada. But despite being a great adventure for many young people, summers in the Arctic could also be a hardship at any age, and the selection was critical.

Applications were invited and received from across the country, but except for neighbouring universities, there was no chance for personal interviews. We relied instead on academic records, personal letters, and letters of reference from faculty members. In university geography departments where I was well acquainted with members of the faculty, we could rely on forthright letters of recommendation: thus, Ross Mackay (UBC), Brian Bird (McGill), Louis-Edmond Hamelin (Laval), Marie Sanderson (Windsor), and Bill Birch (Toronto) could inform potential applicants that a summer in Baffin Island could be very strenuous and should never be regarded as a holiday. Alternative openings were available, for office appointments in Ottawa. Nevertheless, recruitment was undertaken as a fair and open competition, with the exception that, at this stage, women could not be assigned to fieldwork. Partly to improve our system of contacts, I had begun to accept invitations to lecture at various departments of geography across the country. The University of Toronto was one of the first.

Dr. Nicholson, to whom I invariably presented my initial field plans (although he was more interested in the overall scope than the detail), expressed concern for the safety of inexperienced undergrads in isolated and mountainous Arctic terrain. I assuaged his fears as best I could. Group meetings were set up for the selected students, in which to outline expected behaviour (especially how to react in the event of accidents), first aid training, and other safety factors. As the size and complexity of the Baffin operation grew and fieldwork extended into the fiords of the northeast coast (polar bear territory), we accepted an offer by the RCMP to provide instruction in the use of firearms, principally high-calibre rifles—equipment that, to my relief, was never needed. On a more formal footing, everyone, student assistants and regular staff alike, underwent a fairly rigorous medical examination prior to departure for the field.

For the 1962 summer expedition, Professor Bill Birch (University of Toronto) had provided one of the strongest recommendations I had read to date. Nineteen-year-old Michael Church had outstanding grades and easily passed the branch selection process. He duly reported for his medical examination with the others. The following day, however, I received a telephone call from the examining medical officer. Mike had been rejected as unfit for fieldwork in the Arctic because he was significantly underweight. I was shocked—and so was Mike when I invited him into my office to tell him the bad news. He was aghast and protested; after all, he took part in athletics, was an enthusiastic back-country hiker, had never been seriously ill, and so on. I was interested in his general activities and especially in his high grades in mathematics. I told him I would do my best to overturn the decision. I got in touch with the medical officer and repeated what I had learned from Mike, concluding that I wanted him in Baffin Island. I was informed that there was only one recourse: the federal government would accept no responsibility. Surely, I was told, there are plenty of other students available. My response was an emphatic "No!" So, against the advice of the medical officer, I signed a formal document accepting full personal responsibility, including the costs of any emergency evacuation, hospital expenses, and insurance that his family might choose not to carry. In other words, I decided to take the risk—though Dr. Nicholson told me I was an idiot. I like to think that I made a small but important contribution to this very young man's career.[9] Mike's performance in Baffin Island was exemplary, that season and in many subsequent years.

Preparations for Baffin 1962

The aircraft charter of the previous summer, as related in chapter 2, had been a near disaster, so the selection from competing bids for 1962 was of major concern. In this I was assisted by Murray Sutherland, from our head office, and the process worked out extremely well. The acquisition of specialized glaciological instruments was another challenge, especially in view of Brian Sagar's and Gunnar Østrem's recruitment shortly before the field season. However, the Department of Transport, Meteorological Branch, lent us many of the meteorological instruments, others were purchased new, and Gunnar brought with him from Sweden and Norway a treasure trove of instruments and equipment.

I was able to capitalize on one remarkable success of the previous summer—the special air photography of the entire Barnes Ice Cap and surrounding terrain. The photography had been possible thanks to a personal intervention by Dr. van Steenburgh; he had negotiated the assistance of the RCAF 408 (Photo) Lancaster Squadron, based at Rockcliffe, which was primarily responsible for the aerial mapping of Canada.[10] I also needed his support to secure the cooperation of the Surveys and Mapping Branch in order to use the new air photographs in the production of high-quality topographic maps.[11] While it was 1967 before the maps were printed, Sam Gamble, the branch director, was most amenable to setting the production process in motion. The maps were a remarkable resource for subsequent fieldwork in such an isolated area and represented a special measure of Dr. van Steenburgh's support.

Brian Sagar was set to begin glaciological-meteorological studies on the crest of the Barnes Ice Cap. Seismic and gravity studies were added so that the thickness of the ice cap could be determined. Dr. Hans "Housi" Weber (J. R. Weber) had been the gravimetrist and a member of the Swiss mountaineering section of the AINA 1953 expedition to the Penny Ice Cap, in southeastern Baffin Island. He now had a permanent position with the department's Dominion Observatory Branch. Like the Geographical Branch, the Observatory Branch was a very small departmental unit and its director, Dr. Beals, became a most helpful senior colleague. He once explained over coffee in his office that "we small branches have to stick together for mutual support to help avoid being overridden by the 'big three' [GSC, Mines Branch, and Surveys and Mapping Branch]."

In this manner, a joint research effort developed. Housi, a personal friend, was keen to go back to Baffin Island, so we were able to put together a team that led to a much more ambitious glaciological program than I had originally thought possible. My invitation to the GSC to add a Pleistocene geologist to our party was met "with regrets"—there was no available geologist. I sent a similar invitation each subsequent year and always received the same answer. The GSC apparently could not accept the notion that geographers could actually assume the leadership of Arctic field expeditions.

By the close of 1962, everything seemed in place for a long-term study of north-central Baffin Island. The results from the first two seasons had been substantial. The progressive enlargement of the branch staff and rapidly increasing budget augured well for development of a fully interdisciplinary and international endeavour.

CHAPTER 4

BAFFIN 1962

Ice Mining on the Barnes Ice Cap

Brian Sagar, with summer student assistants Uwe Embacher and Chris Bridge, reached Fox-2 via Fox-Main, the centre of eastern DEW Line operations at Hall Beach on the Melville Peninsula, by May 5, 1962. Here a DC-3 on ski-wheels awaited them, although it was not until May 15, after repeated attempts thwarted by fog, that they reached their chosen base camp site on the crest of the Barnes Ice Cap. By May 18, the camp was complete and, as Brian wrote in his field report,

> A 15-knot wind and a temperature of –5°F [–20°C] effectively reduced any tendency to stand and stare. A double-walled pyramid tent was pitched and a hot brew soon ready. Although the sun was always above the horizon, the land was still in the grip of winter.

The work began immediately. The 1961–1962 accumulation season had produced about a metre of snow that lay directly on hard glacier ice of the previous summer, conditions very similar to those determined for the south dome of the ice cap by the 1950 AINA expedition. On May 21, Brian's party was joined by the Dominion Observatory gravity survey team, led by Dr. Housi Weber.[1] The gravimetrists brought in with them a period of clear and often calm weather that prevailed for ten days. A Parcoll hut—a transportable, easy-to-erect, five-metre-wide insulated shelter with an aluminum frame and a wood floor—was erected as the combined team now numbered ten.

Measuring energy balance on the ice cap

Snowmobiles of early vintage, skis, and helicopter transport were all employed in a gravity and snow survey across the waist of the ice cap, together with a snow survey along its entire north-northwest-to-south-southeast length of 145 kilometres. The snow survey was the first step in an attempt to determine whether the ice cap was gaining or losing mass. Housi later reported from his gravimetric data a maximum ice cap thickness of about 620 metres. Brian's team extracted ice cores with a manual ice drill to depths of up to 20 metres. Stakes were drilled in across the length and breadth of the ice cap in order to establish a network for measurement of the progressive snow and ice melt during the 1962 ablation season and the accumulation of the 1962–1963 winter. Indeed, determination of the following winter's accumulation would be one of the objectives for the 1963 "summer."

The glaciologists paid special attention to the energy balance of the ice cap. The measurement of incoming solar and outgoing long-wave radiation required delicate calibration of an array of instruments. That, together with the results from the stake network, would aid in the construction of a preliminary mass balance of the entire ice cap.

As the season progressed, the camp area became a swamp of melting snow. During July, much of the wet snow gradually disappeared, leaving a surface of loose crystals of melting ice. The team enjoyed another period of still, clear weather—perfect for sunbathing. As Brian recalled, "For four whole days, boots were the only necessary article of clothing. By July 20 the margins of many of the remaining snow patches showed a brilliant red algal bloom ['pink snow']." The occasional caribou surprised them, having wandered more than thirty kilometres from the edge of the ice cap, and spectacular flights of ducks skimmed the surface, which was dissected by innumerable meltwater streams.

By late July, all the previous winter's snow had disappeared, along with the refrozen meltwater, meaning that 1962 was a very unhealthy year for the ice cap. However, the onset of winter was rapid. A blizzard accompanied by heavy icing brought down the radio masts in early August. By mid-August, there was a surface of firm snow ideal for ski travel. It was time to head back to Ottawa. Equipment and surplus food were stored in the Parcoll hut for the next summer. Another fall of snow made snowmobile travel feasible, so evacuation was effected in comparative comfort, although deep meltwater canyons and poor visibility impeded progress. On August 24, the party reached a prearranged location on the northwestern edge of the ice cap where various meltstreams combine to form the upper course of the King River,

a major tributary of the Isortoq that flows westward into Foxe Basin. This lay only a short distance upstream of the site on King River to which the main base of the 1962 "land party" had to be moved due to a sudden fall in the level of Flitaway Lake.

Geomorphological research from Flitaway Lake and the Isortoq River

The land party operated quite separately from Brian's group on the ice cap until the final days of evacuation in late August. The majority of this group, under the supervision of John Andrews, now a full member of the branch staff, had been flown in to Flitaway Lake via Fox-2 in early June. John's chief assistant was Bruce Smithson, a recent Carleton University graduate student. The Dominion Observatory's chartered helicopter ferried John's two-man kayak and sundry camping equipment to the Isotoq River, thereby saving him a heavy backpack. The main task was a detailed study of the till fabric of the cross-valley moraines in the Isortoq and Rimrock valleys and the complex of moraines and glacio-fluvial features between the terminus of the Lewis Glacier and the confluence of the Lewis and Isortoq rivers. This concentrated work led to a fairly complete understanding of the mode of formation of the cross-valley moraines and several significant publications (Andrews & Smithson, 1966). Later in the season the group was moved several times along Grant-Suttie Bay and Isortoq Lake, eventually being returned to King Lake when they assisted with Gunnar's excavation of a large sample of ice from the margins of the Barnes Ice Cap farther south. During their traverse of the middle Isortoq valley, they discovered extensive sediment deposits rich in plant fossils. Laboratory examination and radiocarbon dating from beyond the limits of the method demonstrated that an unusually important find had been made: proof of the existence of a rich vegetation cover. (The significance of this discovery is discussed in more detail in chapter 11.)

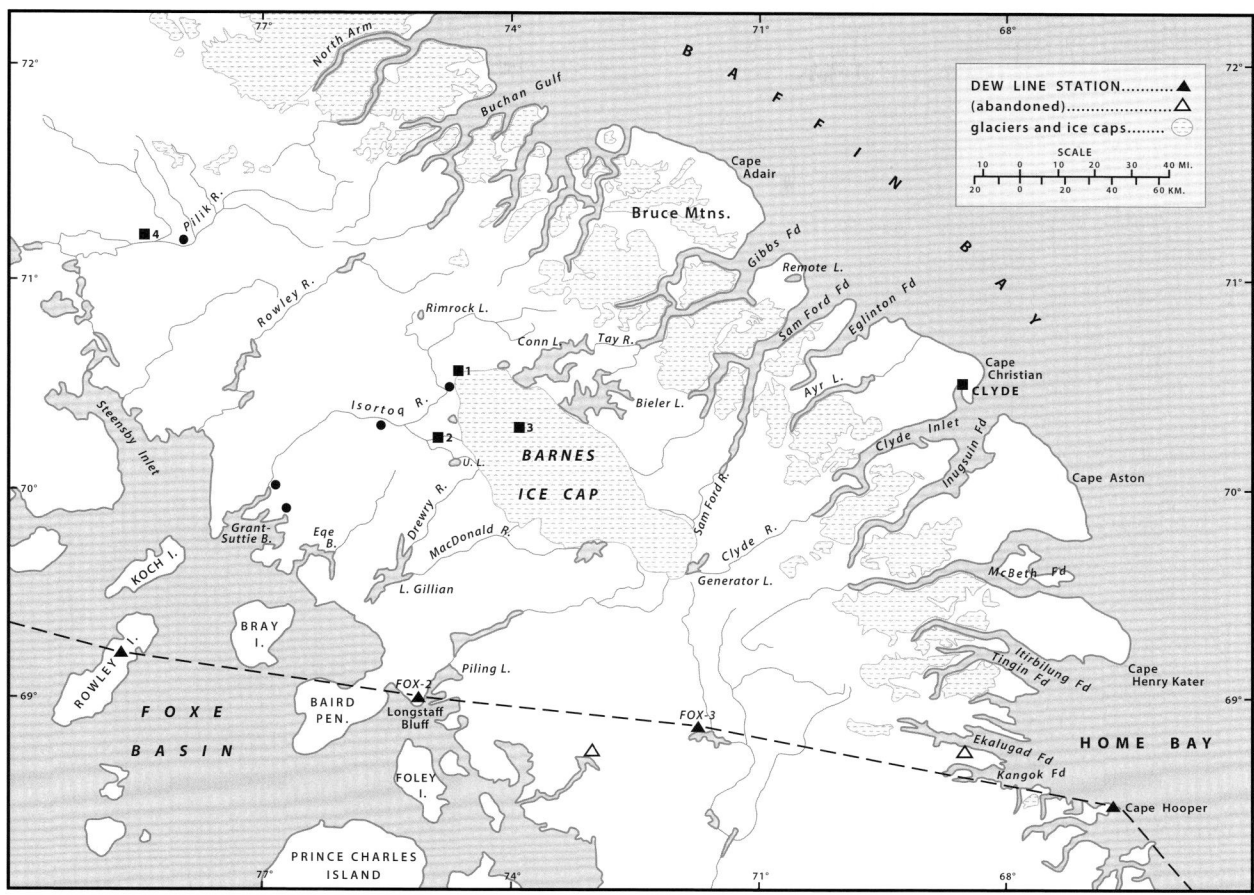

MAP 5: The base camps (black squares) and secondary camps (black dots) used during 1962: (1) Flitaway Lake, (2) King River, (3) Barnes Ice Cap crest, and (4) Pilik Lake. (UL represents Umbilicaria Lake.)

George Falconer, with Mike Church as his student assistant, reached Flitaway Lake by the now standard route via Fox-2. This was very timely as they also could be assisted by the Dominion Observatory's helicopter. They spent the summer on Pilik Lake, more than one hundred kilometres north of Rimrock Lake. Unbeknownst to George and Mike at the time, the fabled Arctic prospector Murray Watts was involved in preliminary mineral exploration in the general vicinity. On August 1, with his Cessna pilot, Ron Sheardown, Murray made a surprise

"coffee-stop" visit. Again, nine days later, Murray and Ron dropped in on George and Mike as the latter were working out of a fly camp that was a good day's hike from their base camp. This netted them a flight back to their base camp. George did not realize until later that Murray had surreptitiously "borrowed" from his collection of air photographs one that turned out to serve a rather important commercial purpose. Nevertheless, this Arctic wilderness acquaintance led to very useful assistance the following year.[2]

Scandinavian contacts and northern travels with Gunnar Østrem

My personal involvement in the Baffin Island research was delayed as I had been invited to attend an international symposium in Reykjavik, Iceland, in mid-July. The symposium was entitled "North Atlantic Biota and Their History" (Löve & Löve, 1963). I presented a paper on the early Baffin Island results (Ives, 1963), leading to valuable international exposure and constructive criticism. Many of the leading Scandinavian biologists, geographers, and geologists attended, including professors Gunnar Hoppe and Eilif Dahl, whose research had greatly influenced my own. A further advantage was that I persuaded Ross Mackay, who had been spending a sabbatical year from UBC with us at the branch, to make his first visit to Iceland with me.

The Iceland venture would significantly delay my departure from Ottawa for Baffin. However, this enabled me to travel north with Gunnar Østrem, whose appointment was completed, so that he would arrive in Ottawa within a few days of my return from Reykjavik. Although the field season was half over before we left Ottawa, the experience of travelling with Gunnar as our newest member of staff was worth it.

Gunnar and I took commercial flights on August 1 from Ottawa via Quebec City to Sept-Îles, on the north shore of the Gulf of St. Lawrence. Here we met with the staff of the bush charter operator, Northern Wings, from whom we had arranged to charter a Beaver on pontoons to fly us all the way to Fox-2. We spent a comfortable night and the next morning had thorough discussions with the Northern Wings staff, regaling them with the horror stories of the Cessna charter of 1961. They took our point of view very seriously, assuring me that Spike Burnett, who was to be our pilot, was the company's senior flyer and that we would also have a fully trained mechanic along with us. Spike, one of the ultimate "characters" of the northern bush pilot fraternity, became a central figure in this next phase of the Baffin Island project.

The flight to Fort Chimo was uneventful, though it refreshed my interest in the vast territory that I always refer to as Labrador-Ungava (Ives, 2010). The first leg of the journey involved an overnight at Schefferville, my former home from 1957 to 1960. Here we were guests of Joan, wife of Bill Mattox, who had succeeded me as field director of the McGill Lab in 1960. I had unexpectedly met Bill in Reykjavik a few weeks earlier, where he had been awaiting a flight to Greenland to pursue his main research objective, the Greenland Falcon. The flight in and out of Schefferville gave me the chance to point out to Gunnar many of the diagnostic glacial features, such as the glacial meltwater drainage channels, that had been studied in the 1950s.

We landed at Fort Chimo in the early evening of August 3. A pal of Spike's met us and took us to the only restaurant for a good supper, and we slept in the Northern Wings camp. We were hoping to be away the next morning after an early breakfast, but were delayed by reports of poor weather in Hudson Strait. In fact, this was a matter more of uncertain weather than of any unreliable weather forecast. We were on the edge of the Arctic, and there were considerable stretches of possibly ice-choked water between Fort Chimo and Frobisher Bay. To make the crossing to Baffin Island in comparative safety, we would need the good weather of a pronounced high pressure system that extended all the way, a distance of more than six hundred kilometres.

The weather on the second morning was no better, so I proposed to Gunnar that we make an aerial reconnaissance around the south and east coast of Ungava Bay. Perhaps we could reach Abloviak Fiord, to the head of which Pauline and I had backpacked in 1956. Then I realized that Olav Løken would be halfway through his third field season in the Torngat Mountains. How magnificent it would be if we could make a surprise visit! Unfortunately, deteriorating weather caused us to turn back a short distance from the entrance to Abloviak Fiord, although we did manage to make landings on three small lakes not far from the coast.

These brief landings were worthwhile as we obtained some key field data relating to the history of Laurentide Ice Sheet recession at the close of the last ice age. Although the main flow of ice had been from Ungava Bay northeast through the Torngat Mountains and into the Labrador Sea, we found evidence for a late movement of ice from the land to the northwest into the bay. Gunnar's presence was valuable as he provided an independent expert check on the ice movement determinations. This was important because of ambiguity about the late-glacial ice movements in relation to the drainage of the Naskaupi and McLean glacial lakes and the final disappearance of the ice sheet from Labrador-Ungava (Ives, 2010), factors related to our Baffin Island research.

Attracting unwanted international attention

After returning to Fort Chimo, we found that the weather remained uncertain. We resorted to a compromise whereby Gunnar and I would take the commercial Nordair flight to Frobisher the following day. This would allow Spike to have two 45-gallon drums of aviation fuel loaded into the Beaver's cabin with a feeder hose and pump so that his mechanic could refuel. We would all meet up in Frobisher, although another four days would pass before Spike was able to make the long crossing. We enjoyed excellent food and sleeping quarters at the US Strategic Air Command (SAC) base in Frobisher while awaiting Spike's arrival. But that otherwise welcome arrival produced a totally unexpected "snafu," landing us in potentially serious trouble with the senior SAC officer.

Despite Spike's successful completion of a challenging flight, this was his first encounter with the problems of landing a floatplane at Frobisher. The bay experiences a very large tidal range that exposes a wide rocky seashore at low tide. Spike certainly knew the problems associated with such landing conditions but was not aware that it was normal practice for Frobisher. Instead, spotting what appeared to be an eminently serviceable lake close to the SAC base, he effected a perfect landing—on the reservoir that was the base's drinking water supply, unfortunately. He and the Beaver were immediately taken into custody by US military personnel, thereby causing a rather severe problem. This was classed as an international incident that I was not allowed to discuss with the SAC commanding officer, as DEW Line operations were still "secret." All communications for the release of Spike and his Beaver had to be conducted between Ottawa and Washington, DC. It was remarkable that we were "released" and able to fly on to Fox-2 the following day.

Creating links between lab staff and scientists in the field

Rolf Kihl, our new geomorphology lab technician, had been flown in to Flitaway Lake with John Andrews and the group of "early birds" that included George Falconer, Mike Church, and Bruce Smithson. Rolf's task was to set up a geomorphological field laboratory for processing the large number of soil samples that I expected the field parties would collect. This lab would allow Rolf to complete the bulk of the grain-size analyses and provide preliminary results for the field staff immediately. It also represented my first attempt to give branch support staff an opportunity for direct involvement with the field crews. In this way, I hoped that the experience would contribute to a fuller understanding of the critical link between field research and the more mundane desk and laboratory work in Ottawa.

Although Spike and the Beaver were available for only a limited time, we accomplished a great amount of fieldwork, including inspection of the progress of the entire expedition, logistical support, and valuable long-distance reconnaissance. I was able to visit George and Mike on Pilik Lake, and I moved them to Tay Sound on August 18. This enabled George to survey the upper limit of late-glacial marine activity and to make a critical collection of seashells that dated this at 8,350 years before present. Spike managed to retrieve the two of them five days later in marginal flying conditions and return them to join the main group at King River.

Having dropped off George with the King River party, Spike flew Mike Church and Bruce Smithson to Fox-2 that same day (August 23), thus beginning the end-of-season general evacuation process. Also, as part of this process, George went upriver to the margin of the Barnes Ice Cap to meet Brian Sagar and party. He then accompanied them to the King River base as the first step in their journey back to Ottawa.

Reconnaissance to the eastern fiords and the Cockburn Moraines

Amid all these logistics, I took the opportunity for a long day's reconnaissance to the northeast coast. I had been unable to examine the Cockburn Moraines or to reach the coast in 1961 because of the mechanical problems with the chartered Cessna (see chapter 2). Now I had a first class opportunity. I asked Spike to fly Gunnar and me through the Bruce Mountains to the outer coast, close to Cape Adair. During the flight, we were impressed by the grandeur of the mountains and fiords, small ice caps, and valley glaciers. We made two brief landings and were able to walk out to the coast and record the height of the uppermost raised marine shore features. The maximum height, measured by altimeter, was close to 60 metres, much lower than the 375 metres accepted by Professor R. F. Flint and displayed on his 1945 *Glacial Map of North America*. Farther inland, we observed many instances where local glaciers had advanced down the valley sides to cut through the Cockburn Moraines after the main Cockburn outlet

glaciers of the continental ice sheet had retreated onto the central plateau.

The reconnaissance to the northeast coast reassured me that it had been correct to plan for the eventual extension of the overall project to include a large section of the coastal mountains and the outer coastal lowland. In my early discussions with Dr. van Steenburgh, I had suggested the possibility of a ten-year operation. This second summer made me think, in some ways, that we were likely facing a lifetime of concerted effort.[3]

Moving the Barnes Ice Cap to Ottawa

A major objective of my journey north with Gunnar that summer was to examine the end moraines along the margin of the Barnes Ice Cap and to see if we could obtain a large sample of buried ice to take back to Ottawa. (Figs. 13 and 14) We had pre-selected a section of high moraine from air photographs before leaving Ottawa. The choice was influenced by access to a lake where the Beaver could land, so that the ice sample could be transported back to King River. The most suitable was a small lake (later named Umbilicaria Lake, from one of the lichen species found there during a survey by Pat Webber and John Andrews the following year) situated about fifty-five kilometres south of Flitaway Lake and thirty kilometres southeast of the King River camp. Gunnar was able to sample ice from the ice-cored moraine and make a microscopic comparison with a sample from the nearby ice cap. He had no doubt that the contrast between the two samples—large crystals from the ice cap and very small ones from the ice-cored moraine—followed the pattern of his results in Swedish Lappland. The moraine ice core was not glacier ice but had been derived from wind-accumulated snow subsequently buried and compressed by the ice cap's discharge of morainic debris (Østrem, 1964).

Our next step was to set up an "ice mining" operation. We had brought with us a portable Swedish

Fig. 13: Northeast margin of Barnes Ice Cap. Looking northwestward, showing ice-dammed lakes. The linear features in the foreground are the now largely dry channels cut by meltwater draining along the ice cap edge as it slowly retreated. (Photo: August 1966)

Fig. 14: Southwest margin of Barnes Ice Cap. Running across the midsection of the photograph is a large multiple ridge of closely spaced end moraines. The massive ridge complex is cored by ice. The acquisition of a large ice sample from such a situation was one of Gunnar Østrem's major objectives in 1962. Note the many small gorges on the slope of the ice cap that have been cut by meltwater from the previous winter and spring snow. (Photo: August 1962)

Atlas Copco pneumatic drill ("portable," even though with rods it weighed more than sixty kilograms!) and a score of large plastic bottles. We assembled seven of the group available at the King River camp for hard labour: John, Rolf, Brian Thompson, Claude, Bruce, Gunnar, and me. With Gunnar acting as the "mine foreman," we succeeded in digging out enough ice from deep within the moraine to fill eighteen of the large plastic bottles. (Fig. 15) We backpacked them, together with the portable drill, across more than a kilometre of wretchedly unstable boulders to Umbilicaria Lake. From there, Spike airlifted the bottles via

King River to Flitaway Lake, where George assisted Gunnar in carefully labelling and sealing them to be ready for shipment to Fox-2 and Ottawa. Now I realized how accurate had been Britta Østrem's prediction (see chapter 2, note 5). Gunnar was in the triumphant process of moving the Barnes Ice Cap to Ottawa. The first shipment weighed at least four hundred kilograms, in carefully sealed bottles, and it was not to be the last. Once unloaded, the melted ice would be fully filtered and evaporated so that any pollen spores could be extracted, together with any other carbonaceous material for radiocarbon dating.

Fig. 15:
The laborious process of extracting 450 kilograms of ice from an ice-cored moraine of the Barnes Ice Cap. From left: Gunnar Østrem, John Andrews, Rolf Kihl. (Photo: August 1962)

Spike grunted that he would be branded as the first crazy bush pilot to help transport ice, so very expensively, from the Arctic to Ottawa, when most sensible people would merely wait a few months for the Ottawa River to freeze over and then cut out all they needed.[4] (Fig. 16)

Gunnar presented me with another logistical problem. He needed to get two samples of ice, one from the moraine and the other from the ice cap, for laboratory examination and crystal photography. While these samples were small (weighing less than a few grams each), they had to be prevented from melting and there was insufficient dry ice to achieve that during the long journey south. I was rapidly learning that Gunnar was one of the most persuasive characters I had ever met. Even then, I was amazed to see on the commercial flight from Fox-Main to Montreal

the way he induced the chief flight attendant to allow him (and us) to eat the airplane's entire supply of ice cream so that the precious ice samples could be stored in the plane's small refrigerator. Repeating this tactic when we changed to a Canadian Airlines flight in Montreal, and again on his way back to Stockholm, he gained an "ice cream" reputation that stuck to him for the rest of his life.[5]

Following our return to Flitaway Lake, and because we were approaching the end of the field season, Spike began making the occasional flight (as the weather and local requirements permitted) between Flitaway and King lakes and Fox-2, as the first stage of transferring equipment and field samples due for shipment to Ottawa. I accompanied him on one of these flights and thereby joined him in a rather startling experience.

Fig. 16: The "ice miners" on Umbilicaria Lake, ready to load ice samples into the Beaver for transport to Flitaway Lake and then south to Ottawa. From left: Brian Thompson, Rolf Kihl, Gunnar Østrem, Claude Lamothe, John Andrews, Bruce Smithson. (Photo: August 1962)

A tale of floatplane payloads and punctures

Spike Burnett and Northern Wings dominated our open-season logistics from 1962 to 1964, and there are many tales to tell. This one is fairly typical. Spike was, above all, a jovial fellow, but with a tendency to gruffness if provoked. He was well liked by students and staff. I quickly developed a cordial working relationship with him. I was party leader, as he liked to say, but he was commander of the "air force." Understandably, I always wanted maximum Beaver payloads; appropriately, Spike insisted on caution. I well remember this occasion as we were in the early phase of evacuating the Flitaway and King River camps to Fox-2. As we inspected the load that I wanted taken from Flitaway Lake to Fox-2, Spike adamantly refused to take the last two pieces from the

pile we had assembled. In his trademark gruff voice, speaking now decidedly from his position of superiority as the Northern Wings senior pilot, he muttered, "Jack, I have known many bold pilots, I have known many old pilots, but I have never known any old bold pilots!" So, of course, I demurred. (Fig. 17)

That early morning, a small part of the ice cliff that dammed the lake collapsed into the water and many pieces of ice floated out across the lake. Our cautious "old pilot" had been eyeing them dubiously all morning and well into the afternoon. After lunch, however, a light breeze sprung up and succeeded in marshalling them all into a tight corner beneath the ice cliffs. So, as was usual on Flitaway, we climbed on board and backed up the Beaver until the tailplane was almost over the lakeshore to give us maximum length of water for takeoff. This was necessary because the distance for takeoff was critically short.

Fig. 17: Beaver aircraft landing on Flitaway Lake. A perfect performance by Spike Burnett on a mirror-like surface. The scattered pieces of floating ice help a pilot to estimate his height above the surface—provided he doesn't hit one! Barnes Ice Cap in background. (Photo: August 1962)

Spike's standard drill in this kind of situation was to make sure the engine was well warmed up and then open the throttle rapidly, rock the plane as soon as we had reached enough speed, and seemingly jerk it up off the lake surface. This time, just as he was in his final act of liftoff, we heavily clipped a stray piece of almost totally submerged ice that had gone undetected. The plane lurched abruptly and I was afraid that we would do a nosedive into the lake. Spike expertly rose to the challenge and kept us in the air. We circled very low across the gently sloping ground beyond the shore and slowly climbed, turning south toward Fox-2. Spike was silent for some minutes as the Lewis Glacier slipped away beneath us and we approached the King River. I could see that his face was glistening with perspiration. He turned to me

and shouted above the roar of the engine, "If we had taken those last bloody boxes of yours, we would have been in the drink."

We continued gaining altitude for several more minutes. There was something decidedly odd about our angle of flight; it was as if one wing was longer than the other and causing our alignment to be about twenty degrees off our line of flight. "So you have noticed it," said Spike, more gruffly than usual. He explained that we must have torn a piece out of the port-side pontoon, resulting in air drag that made us fly in a crab-like manner. He added, "We're in for trouble when we try to land at Longstaff [Fox-2] as I expect we will sink!" The prospect was an alarmingly chilly one, especially for a non-swimmer.

Spike explained that he would try to land as close to the shore as possible as there was a narrow stretch of shoal water along the edge of the lake. If we didn't make it to the shallows [shoal water], real trouble would indeed ensue; if we were too "successful," though, we might not have enough water left before crashing into the boulder barricade that fringes the lake. The landing would require a great deal of finesse with little or no margin for error. So Spike radioed ahead to Fox-2 to alert them to our predicament, requesting that "all possible emergency equipment" be at the lakeshore to meet us.

Some forty minutes later, as we were approaching the lake, Spike instructed me to make sure not only that my seat belt was tight, but that I could release it very quickly. "I assume you can swim?" he asked.

I replied, "No!"

He groaned. "We will have to risk hitting the boulders—damn you Brits, why didn't you learn to swim?" Spike's tirade went on a lot longer, with an unprintably colourful overtone, although I realized that he was simply relieving his own tension. There was really no answer, but I had always been more mindful of the prospect of cardiac arrest than of the possibility of drowning in cold Arctic water. Discretion prompted me to keep this thought to myself.

The sun was dipping toward the horizon as we came in very low along the lake, trailing the heels of our pontoons, with marvellous finesse, lightly in the water to reduce speed. This gave Spike the problem of trying to maintain a fairly straight course despite the punctured pontoon, although he appeared in absolute control. Then we plopped down on the full length of the pontoons and abruptly lost speed. I gasped, realizing that we were less than a few short plane-lengths from the line of boulders that bordered the lake. The Beaver began to tilt alarmingly as we both scrambled out onto the intact starboard pontoon. We

sank in about a metre and a half of water, with the starboard side remaining on the surface. We waded ashore greeted by a cheering crowd of DEW Liners who had assembled in response to Spike's radio SOS. We easily reached dry land, my cameras clear of the water. But Spike was furious. He detested getting his feet wet. He swore that after this he would have his mechanic fly with him at all times. This was the first of many bush plane stories that eventually, and regrettably, resulted in a change of charter company from 1965 onward (see chapter 5).

Our plight was attended to with a speed and efficiency that astonished me. The Beaver was hauled out of the water, repaired, and able to return us to Flitaway within twenty-four hours. It was indeed fortunate that Fox-2 had been our destination, with its availability of cranes, bulldozers, and mechanics. The only "penalty" we had to endure after the enthusiastic reception of the DEW Line crew was a very comfortable night's sleep at the station, steaks for dinner, and an evening of billiards. With my cautious provision of a couple of large whiskies, Spike's mood changed from one of an affronted self-image to one of derring-do, and he was back to acting the part of a remarkably skilful bush pilot—which he undoubtedly was.

By August 30, the entire field crew was back at Fox-2. Spike took off for Frobisher and Sept-Îles via Fort Chimo, and the next day the rest of us travelled by the so-called DEW Line lateral flights to Fox-Main and were soon safely back in Ottawa. Gunnar used the next few days to meet additional members of the branch staff and staff of the National Research Council. In Montreal, he met with Dr. Svenn Orvig and Dr. Fritz Müller, the latter newly returned from Axel Heiberg Island. Gunnar flew back to Stockholm on September 7, determined to return for another assault on the Barnes Ice Cap the following spring.

CHAPTER 5

❧ EXPANDING BAFFIN RESEARCH ❧ AND WIDER RECONNAISSANCE, 1963

After two full summers in Baffin Island, there was a great amount of data to work up, group discussions to be arranged, and decisions to be made concerning future plans. The number of permanent branch staff would have to be increased, and additional able and energetic undergraduate summer assistants would have to be recruited. It was apparent that at least one more season of concentrated work around the northwest end of the Barnes Ice Cap would provide very useful results and that the glaciological work on the ice cap should be expanded. In addition, plans were needed to extend the fieldwork into the eastern mountains and fiords as far as the Baffin Bay coast. This would require not only more staff but also an appreciable increase in our operational budget. It also seemed logical to pursue collaboration with other institutions, and especially university specialists, so as to ensure intensification of the interdisciplinary activities.

Planning for lichenometric studies and glaciology fieldwork

For several years, I had maintained a regular correspondence with Dr. Roland Beschel, a botanist/lichenologist who had earned his DPhil at the University of Innsbruck, Austria, in plant ecology and physical geography. Dr. Erling Porsild of the Natural History Museum of Canada, who chaired the AINA Banting Research Fund, had originally put us in touch when we were both applying for Arctic research grants during the 1955–1956 winter. Roland, then a new immigrant, was on the faculty at Mount Allison University. He had sent me the early papers on his imaginative use of the growth rates of rock lichens as a means of dating rock surfaces, especially glacial moraines in the Austrian Tyrol (Beschel, 1957, 1961).[1] Following his suggestion, I had visited the British Museum in 1957 to examine rock lichen specimens, principally those of the *Rhizocarpon geographicum* group, members of which are especially slow-growing and long-living (i.e., over a thousand years). In London, I had also been able to discuss problems of species identification with the museum's curator. In 1960, however, Roland had accepted a faculty position in the biology department at Queen's University, in Kingston, so visits became much easier.

Roland was a fascinating and likeable, although very intense, personality and was a welcome guest at our new home, close to the Geographical Branch in Ottawa. He showed great interest in my discovery of the "lichen-free"[2] areas north of the Barnes Ice Cap and agreed to come north with me to validate the tentative conclusions that I had made. This in turn led to his suggestion that we add to our 1963 field party one of his graduate students, Patrick (Pat) J. Webber, who had just completed his master's degree and knew how to identify lichens and work with fossil plant material. Roland proposed that Pat could then undertake plant ecological studies for his doctoral research. He would also be able to help us expand our use of rock lichens for dating the retreat phases of the Barnes Ice Cap—in all probability over a period of more than a thousand years. As a quid pro quo, such an arrangement would justify Roland's visit, both to inspect Pat's progress and to accompany me to Rimrock Lake to cross-check the interpretation of the "lichen-free" areas.[3] In ways such as these, the inclusion of faculty and graduate students from Canadian and European universities, as well as from disciplines other than geography, began to expand.

It was necessary to coordinate Gunnar Østrem's logistical complexities, as he was still straddling overlapping appointments in Ottawa and Stockholm, with his family living in Stockholm. Plans were laid for Gunnar to arrive in Ottawa at the beginning of May 1963 and accompany the early field groups to Flitaway Lake and the Barnes Ice Cap. His main task would be to initiate studies in glacio-hydrology on the Lewis Glacier and Lewis River and to train Canadian undergraduate field assistants to maintain detailed field observations throughout the season.

The next component of the 1963 season would be a series of studies to expand our understanding of the cross-valley moraines, easily accessible from Flitaway Lake. Another objective was to examine the relationship between the retreat of the proto–Barnes Ice Cap from the Foxe Basin coast and the timing of the entry of the sea—one of the closing phases of the last ice age. This work would be headed by John Andrews, aided by a small army of student assistants, and would incorporate Pat Webber. Under Roland's guidance, Pat and John could begin to apply Roland's methods in lichenometry to dating the retreat phases of the Barnes Ice Cap. One of the permanent indicators of

this process today is the significant number of place names on the topographic maps of the area that are devoted to species of lichen, including Alectoria Lake, Umbilicaria Lake, and Arenaria River.

Breaking new ground: Women in Canadian Arctic fieldwork

Shortly after my return to Ottawa from Baffin Island in early September 1962, I received the now familiar invitation for "debriefing" from Dr. van Steenburgh. These meetings ranked among the most fascinating and instructive (even inspirational) of all my experiences in the department. On this occasion I was asked to outline at some length what I thought had been achieved so far in Baffin Island and what my plans were for the future. Dr. Van, as he was usually known to the senior staff, asked me many penetrating questions. He was especially interested in my account of the "lichen-free" areas; he posited that I might have hit upon a very interesting new phenomenon. When I mentioned my connection with Roland Beschel, he strongly supported the relationship.

I raised the possibility of helicopter support. Dr. Van advised me to proceed with caution, as a significant increase in expenditure would be needed. He suggested that perhaps by the end of the next field season, I would be in a position to request an augmented budget, but that to do so would require convincing justification. If I could manage that, I would have his full support. He was also interested in the trend toward involvement of university faculty and graduate students from disciplines other than geography. This he warmly supported, but again he advised that an explicit justification would be needed.

After this much encouragement, I decided to tempt fate: "Dr. Van, what do you think about my adding women to the Baffin Island expeditions?"

That produced an emphatic response. "Why not? But you will have very strong opposition, especially from Bob Code [director of personnel] and the GSC.

Give it a try . . . but again, you must make a convincing justification. The reason you give *cannot* simply be a plea for equality of treatment for women. You must have a specific reason."

Therein lay the kernel of my campaign to break down the veritable bureaucratic brick wall and open our Baffin field operations to women, especially undergraduate students. The specific justification would centre around Dr. Cuchlaine King. She had been my undergraduate tutor at the University of Nottingham and a member of two student glaciological expeditions that I had led to southeast Iceland. I wrote to her to see if she could be tempted to come from England for at least one field season on Baffin Island. She did not hesitate, although her arrival was delayed until 1965.

The 1963 Baffin Island field season

The scale of operations in 1963 was increased significantly, with a total of seventeen participants, not including two aircraft pilots and an engineer. Following the early season activities supported by a ski-wheel Otter chartered from Wheeler Airlines based at Frobisher, two Beaver floatplanes were chartered from Northern Wings for the open-water period, to be flown by our old friend Spike and another pilot, Jim Cole. (Map 6)

As in 1962, distinct lines of activity were scheduled. Brian Sagar, assisted by Rolf Kihl, Chris Bridge, Pierre Gaudreau, and G. Emery, was again based on the crest of the ice cap. Gunnar Østrem, with Mike Church and Bill Rannie, was scheduled to begin glacio-hydrological studies on the Lewis Glacier and Lewis River. George Falconer and assistant Kent Sedgewick were flown in to the Baffinland Iron Mines camp at Mary River, close to Pilik Lake. They were invited by Murray Watts to use the company's base camp, a most satisfactory result of the previous summer's surprise encounter.

The largest group, under John Andrews, with Pat Webber, was to focus on the geomorphology, glacio-hydrology, and botany and lichenometry of the area around the northwestern margins of the ice cap, along with the interrelations between late-glacial sea levels and the retreat of the proto–Barnes Ice Cap from the Foxe Basin coast. Flitaway Lake was again to be the main base of operations, its threat to disappear the previous summer turning out to be an apparent false alarm, although we were to remain on tenterhooks. Eventually, later in the season, a substantial transfer from Flitaway to King River was undertaken. (Fig. 18)

Perhaps the most remarkable feature of the summer was the success achieved by John Andrews and Pat Webber in training the student assistants. The party's research task was to apply the lichenometric techniques to relative dating of the progressive surface exposure following eastward retreat of the western margin of the proto–Barnes Ice Cap, during the final phases of the last ice age to its position in the 1960s. Extensive "old" deposits of fossiliferous peat were located and sampled. In addition, the team completed the precise levelling of the raised marine shore features along the Foxe Basin coast, the potential importance of which had been realized during the 1961 reconnaissance. John's work led to the identification and dating of the marine limit, which he determined represented an actual contemporaneous sea level (or "strandline"; see Løken, 1960). Furthermore, he was able to show that lobes of the remnant Baffin Island ice cap still penetrated to tidewater in the Grant-Suttie Bay area when the late-glacial sea had reached its highest level on the exposed headlands. The regional tilt was up toward the southwest, thus confirming our hypothesis based on the 1961 reconnaissance (Ives & Andrews, 1963). John also confirmed that the age of the associated Isortoq moraines was between 7,000 and 5,500 radiocarbon years. Another notable innovation in Canadian Arctic research was Gunnar's introduction of Norwegian methods for calculating turbulent glacier meltwater runoff. This

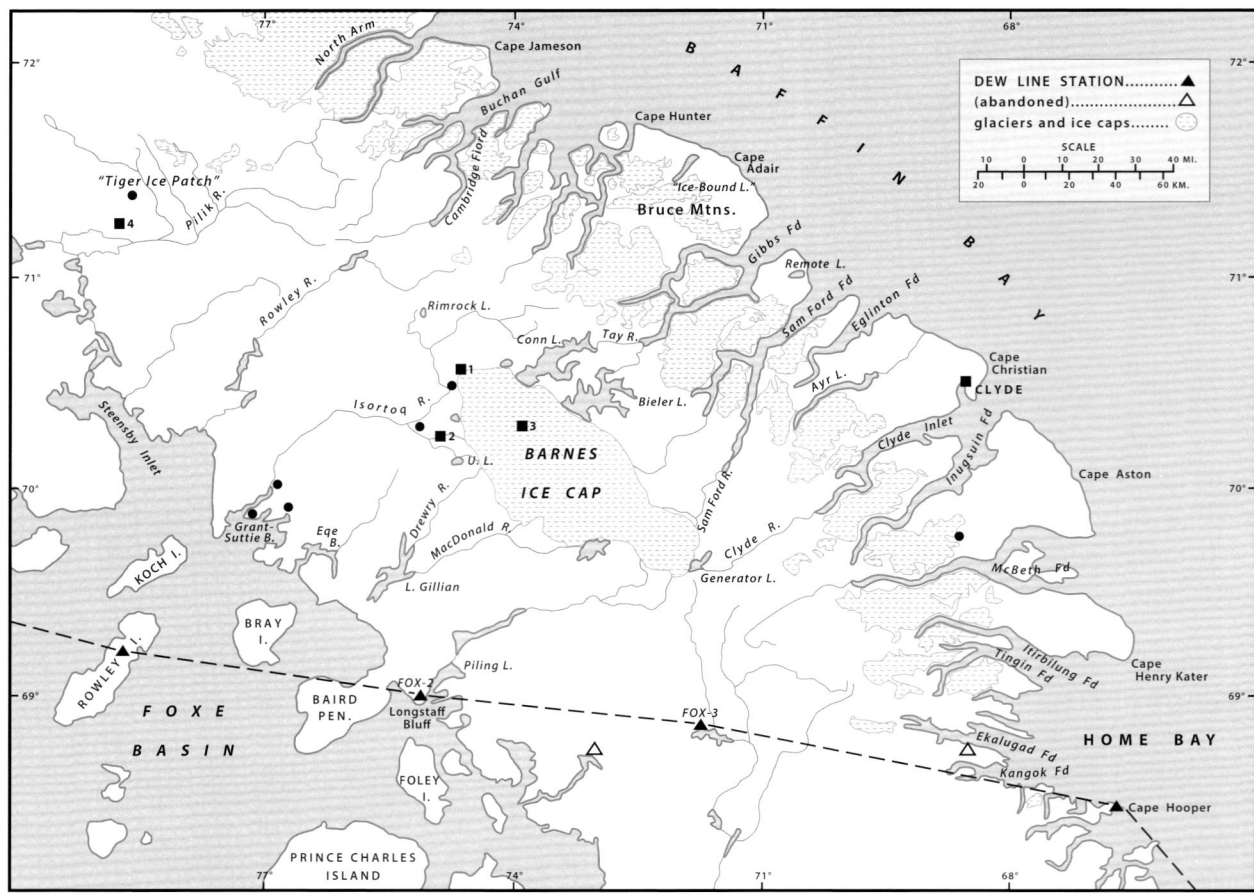

MAP 6: Sites of camps and bases for 1963 (see also Map 7): (1) Flitaway Lake, (2) King River, (3) Barnes Ice Cap crest, and (4) Mary River–Baffinland Iron Mines base. (UL represents Umbilicaria Lake.)

involved training Mike Church and Bill Rannie to take care of the season-long discharge measurements of the Lewis River using the "relative salt dilution" method (Østrem, Bridge, & Rannie, 1967). In practice, the approach required all of Mike's inventiveness to overcome problems inherent in Gunnar's methods, in part resulting from the extreme turbulence of the river. Gunnar was also able to obtain another large ice sample from the Barnes Ice Cap ice-cored moraine close to Umbilicaria Lake, this time early in the season using the ski-equipped Otter.[4]

Brian Sagar, supported by Rolf Kihl, Chris Bridge, G. Emery, and Pierre Goudreau, reoccupied their old camp on the crest of the ice cap in late April.

Using the Otter, snowmobiles, and skis, they inserted ablation stakes over an extensive area. They operated as a self-contained group on the ice cap until the end of the season, with one exception. Brian made a "breakout" visit to Flitaway Lake, where Gunnar invited him to give a talk on the work in progress to the quite large group then present. The main "land party" left Ottawa on May 23, early enough to use the Otter, although it was into June before the Flitaway base camp became fully operational, owing to bad weather that restricted airborne operations.

During this early period, a curious personal problem caused a degree of consternation and required my intervention from Ottawa. When Gunnar arrived in

Fig. 18: King Lake after the emergency move from Flitaway Lake. In contrast to Flitaway's abrupt fall in level, King Lake has risen, due to the warm weather accelerating melt on the ice cap that supplies most of its water. (Photo: August 1963)

Ottawa, he confessed that he had been exposed to mumps through one of his children. As he had never contracted the disease as a child, this presented a delicate problem for the entire group, especially as treatment for an adult male would involve confinement to a darkened room. This would hardly be possible at 70°N during the summer solstice. I authorized his departure but had to ensure that the chartered Otter be kept on hold for an extra ten days, to await the end of the mumps incubation period. It was an expensive adjustment, and I was very relieved when I heard from Fox-2 that Gunnar had remained in excellent health. I am sure it was also a relief for him and his colleagues in the field.

After a break in Ottawa from his ice cap duties during the first part of the season, Rolf Kihl returned in early July to re-establish the geomorphological lab at Flitaway. Considering the large number of field assistants that John had working on till fabric analysis and overall soil sample collection, Rolf made an important and timely contribution.

From the highly convenient Mary River base, George and Kent received considerable assistance. Ron Sheardown, Murray Watts's "pilot-in-residence," took them on an extensive reconnaissance flight across a large section of the northern Baffin terrain, almost as far as the Inuit settlement at Pond Inlet. They were also able to spend five days working on

Fig. 19: With canoe on Rimrock Lake. Roland Beschel (right) couldn't resist a spin on Rimrock Lake with his doctoral student, Pat Webber (left), after inspecting our rock lichen stations. (Photo: August 1963)

and around what became known as the "Tiger Ice Patch," a day's backpack north of the camp. There they greatly benefited from the cache that the "iron miners" had put down for them the year before. The ice patch had been examined by George on air photographs in Ottawa. It had been immediately apparent that the discussion generated by the initial interpretation of the "lichen-free" areas farther south warranted detailed investigation of an actual field site where a small, thin ice patch had been retreating, at least since the first air photographs had been taken in August 1948. In the process of scrutinizing the ice patch, George discovered mosses and lichens emerging from beneath the receding ice margin. He raised the possibility that such vegetation may have survived dormant, not dead, for a very long period. Recent extensive research has demonstrated that mosses and lichens have survived for at least fifty thousand years. This is now being used as significant evidence in support of the highly relevant conclusion that Baffin Island's summer temperature over the last decade (i.e., post-2000) had been higher than at any time

since at least the onset of the last major glacial period (Falconer, 1966; Miller, Lehman, Refsnider, Southon, & Zhang, 2013; La Farge, Williams, & England, 2013). The significance of this finding is discussed in greater detail in chapter 11.

I spent nearly six weeks in the field myself, from mid-July until the end of August. My purpose was primarily a mission of inspection and a start on eastward exploration to determine the next series of field areas that should be tackled in the mountains and fiords. I was also able to be up north to welcome Roland Beschel, who arrived for the final two weeks of August. During his visit, as consultant to the Geographical Branch, he was to review the level of reliability of the extensive lichenometric work that had been spearheaded by Pat and John and to supervise Pat's identification of a suitable doctoral research project. Roland and Pat also flew with me to Rimrock Lake to examine what had become the type area for determination of the cause of the "lichen-free" pattern that was so extensive across the central and eastern sections of Baffin Island. (Fig. 19)

Lakewater levels and "The Little Beaver"

In early August we had another scare. Once again, Flitaway Lake threatened to live up to its name, suddenly dropping in level by more than a metre. This prompted another precautionary evacuation to King Lake to our south (actually a broad lake-like expansion of the King River that drained directly from the Barnes Ice Cap). The days of August 9 and 10 were spent collapsing the lab and most of the tents and, with the two Beavers, ferrying everything to the much more stable southern site. Here, on a long sandy spit connected to a small island close to the south shore, we built a comfortable new base camp. Fieldwork then settled down to a new routine, if "routine" is the appropriate term. The various field camps were once more reoccupied. Roland accompanied Pat, and our pilot Jim Cole worked round the clock to move the various parties from one campsite to another. Mike Church and Bill Rannie had remained behind at Flitaway so they could complete Gunnar's full program of observations on the discharge of the Lewis River.

I returned with Spike and Roy, his engineer, to the King River camp, where Rolf and Pierre Gaudreau were processing a large number of soil samples in the field laboratory. By this time, Spike had begun to expect a surprising level of luxury. He was insisting that engineer Roy accompany him on every flight, which had the effect of reducing the Beaver's payload by about a hundred kilograms. Much to the open amusement of the students, Spike insisted that Roy piggyback him onto and off the plane at each takeoff and landing. (It was normal practice for everyone to jump off the pontoon into the water when a single leap to dry land was not possible.) Spike's Beaver became known, sarcastically, as "The Little Beaver" on account of its small payload compared with that of its twin, piloted by Jim Cole. (Fig. 20)

Following a smooth landing on King Lake, Spike propelled the Beaver so that its pontoons, as usual, glided gracefully onto the sand spit. We went into the main hut to prepare supper. After supper a radio schedule with Brian Sagar on the crest of the Barnes Ice Cap warned us that, even at our much lower elevation, we should be prepared for our first below-freezing temperatures of the summer. Brian reported a hard freeze on the crest of the ice cap and predicted that, as the surface meltwater around his camp was already frozen, the level of King Lake was liable to drop during the night, because most of its supply of water was melt from the ice cap. That warning led to my first serious confrontation with Spike.

After closing the radio sked with Brian, I suggested to Spike that we should all help turn the Beaver around so it faced out into the lake and then anchor it so that it was afloat. Spike gave a very disgruntled response, protesting that he had done enough work for one day. He also claimed that I didn't know what I was talking about. He had "flown around Labrador for almost twenty years" and was "well aware that lake levels didn't fall until late September or early October." I delicately pointed out that we were more than 1,500 kilometres north of his home territory of Labrador, and furthermore, our lake's supply came from the ice cap and it was freezing in place. Unfortunately, Spike took my response as a challenge to his superior judgement, which did not ease the situation.

Much debate ensued. Hadn't it been agreed that while I was expedition leader, he was air commander? Eventually we were able to reach a compromise. We turned the Beaver so that it faced outward but was still firmly grounded on the beach. With no trees around, in contrast to southern Labrador, one of Spike's fears—and it was a reasonable one—was that a sudden high wind could be disastrous for an inadequately anchored plane. In "normal" circumstances when north of the treeline, it is wise to drive a floatplane's pontoons hard up on a sandy beach. In such a situation, two anchors could be embedded farther up the beach. South of the treeline, a plane could also be roped to any sturdy trees close by. And so to bed; it now approaching late August, we all fell asleep in the darkness.

Fig. 20: "The Little Beaver." Engineer Roy is piggybacking senior pilot Spike Burnett, to ensure that Spike doesn't get his feet wet. King Lake. (Photo: August 1963)

I was up early the next morning as usual. I crept outside to admire the still clear sky—not a ripple on the glassy surface of the lake—but I saw with alarm that its level had fallen noticeably during the night even though there had been no freeze. There stood the Little Beaver—high, dry, and quite some distance from the nearest shallow water of the lake! Smiling grimly to myself, I went back into the hut and began to prepare coffee. When ready, I took a steaming mug over to Spike, who was just beginning to emerge from his sleeping bag.

"What's the matter with you, Jack? Why have you taken on the duties of nursemaid?"

"Because I think you will need this, good and hot," I replied. "I only wish I could top it up with something much stronger."

Spike's response: "What on earth does that mean?"

To which I replied, "Spike, please drink it hot, then go out and take a look at the lake."

Poor Spike! He shot out of his sleeping bag, spilling his coffee, rushed out of the hut, and simply stood there in amazement for a few moments. Rolf, Pierre, and Roy, all still half asleep, were roused by a string of bellowed obscenities, which eventually trailed away into speechlessness. I bit my tongue hard to keep back the almost inevitable "I told you so."

Breakfast was acutely tense. When Pierre eventually set the rest of us going in a bout of laughter, Spike just rushed out of the tent to stare again in silence at the Beaver. Pierre whispered, "Should we ask him where he keeps his spare set of wheels?"

"Don't you dare," I replied, "or you will be in real trouble from me!"

What to do? We began by taking out the large plywood sheets that formed the floor of the Parcoll hut. These, along with pieces of smooth fabric, were placed in front of the Beaver's pontoons. Next we harnessed lengths of rope to both fore and aft pontoon struts, two lengths on each side. Spike crawled into the pilot's seat. Our plan was that the four of us would heave on the ropes as Spike started the engine and partially opened the throttle. With Pierre behind me and Rolf and Roy on the far side, we pulled mightily, inch by inch bringing the Beaver a little closer to flotation. Spike opened the throttle a little more and the Beaver trembled from nose to tail. With a series of little jerks, we inched it closer and closer to the water. Finally, after several much-needed rests and severe strain on ropes and engine, we had the plane in the water—but only very shallow water at that, and it was still some way from where it would float. More pulling, with Spike opening the throttle even more so that I was afraid I would see the plane disintegrate. Eventually I found myself thigh deep in the lake; as we gained partial flotation we realized that our efforts would succeed. Spike signalled me from the cockpit that he was about to try three-quarters throttle.

At that moment an anxious thought raced through my mind. Pierre was just in front of the tailplane. What if Spike suddenly unstuck the Beaver and it shot forward quickly? I was in the act of turning to warn Pierre when it did exactly that. Pierre, fortunately, had had the same thought and had ducked, but I was turning into it. The tailplane clipped me on the temple. I lost consciousness and dropped into the lake. The very cold water brought me to my senses almost immediately to observe the Beaver gently floating and a very anxious Spike looking out to see what damage he had done. All I received from the adventure was a soaking, a large bruise on my temple, and a broken pair of sunglasses. It was time for mid-morning coffee, well deserved by all concerned. Spike expressed profound apologies for knocking me out and nearly drowning me, but I laughed it off and our usual good spirits were finally restored.

Reconnaissance to the northeast coast: Critical moments in logistics

Jim Cole, piloting the second Beaver, rejoined us after an extensive tour of work along the Foxe Basin coast as far north as Steensby Inlet. A spell of fine weather prompted a reconnaissance to the far northeast coast on the edge of Baffin Bay. Since 1961, we had accomplished so much fieldwork in the area between the Barnes Ice Cap and the Foxe Basin coast that it was time to shift focus toward the mountains and fiords in the northeastern coastal region.

Gunnar and I had made a single reconnaissance flight to Cape Adair the previous year, and graduate student Dave Harrison, with Dick Cowan (both had completed master's degrees based on work at the McGill Sub-Arctic Research Lab), had spent two weeks working in a small area of the Bruce Mountains this summer. No other investigations had been attempted since the 1950 AINA expedition to the southern dome of the Barnes Ice Cap and the fiord and mountain area between Clyde Inlet and Sam Ford Fiord. Air photo interpretation during the 1962–1963 winter had made it clear that here was a prime research area relatively close at hand. However, there were significant logistical problems to be overcome and an aerial sortie appeared essential.

A reconnaissance to the northeast coast with Jim and the Beaver would require more fuel than the regular tanks could hold. So we hefted four ten-gallon steel drums of fuel into the rear of the cabin. We assumed that along the way we would be able to find open water where we could land and refuel. On Jim's insistent questioning, I assured him that Spike had landed Gunnar and me on a very suitable lake the previous summer. It lay close to sea level in a wide glacial trough subject to high winds that should have aided in clearing away the previous winter's lake ice and would surely provide a safe landing along the planned route. We would also be flying out there ten days later than last year, when it had been completely open without a single piece of floating ice on its surface. Spike readily confirmed this and gave his approval. As things turned out, however, this proved to be a clear error of judgement, and we were very fortunate not to suffer the consequences.

We packed an emergency tent, sleeping bags, climbing rope, and a goodly pile of ship's biscuits (also known as hard tack, pilot biscuits, or "door knockers"), chocolate, pemmican, and canned sardines, and set off soon after breakfast on August 24. We flew almost due north, passing over the Lewis Glacier and Flitaway Lake in the general direction of the head of Cambridge Fiord. On approaching the fiords, massive stretches of broad ridges came into view. They were the Cockburn Moraines.

On sighting the head of Cambridge Fiord, we changed course to a northwest heading for Coutts Inlet and North Arm, the most northerly of the east coast Baffin fiords before Pond Inlet. The scenery along North Arm was spectacular from our flying height of about six hundred metres above sea level. The rock walls rose high above us, and where valley glaciers flowed down into tidewater, we could clearly see how they had expanded after the main ice age outlet glaciers had receded inland onto the plateau, some thousands of years ago.

We emerged from the fiord in the vicinity of Nova Zembla Island, close to latitude 75° north, and turned due east toward Cape Jameson on the outer coast. There had been absolutely no open water along the flight route so far, except for inadequately small areas where mountain streams discharged into the fiords. The prospects of finding a refuelling spot for landing were appearing bleaker by the moment, but I was still optimistic. Now, as we crossed the coast into Baffin Bay everything appeared solid white, except for a distant glimmer of open water that would have been at least fifteen kilometres out into the bay. Jim nudged me. "Jack, you had better locate your famous lake before too long. Are you sure you can find it?"

"No problem," I replied with confidence. "Before we reach Cape Adair we should be able to turn southwest into a huge deep valley, and then I'll show you ten kilometres of beautiful water."

We crossed Buchan Gulf and sighted Cape Hunter. Jim mentioned again that he was anxious to sight open water as there had been nothing so far but frozen lakes, frozen fiords, and pack-ice-strewn Baffin Bay with its streams of mighty icebergs on their way south. We overflew Cape Hunter. Jim had now become noticeably perturbed: "Jack, where the hell is this bloody lake of yours? We're getting very low on gas."

We were flying along an impressive sheer mountain wall and I turned to Jim, still confident. "You see that big opening ahead; that's the entrance to the valley. You'll see the lake in a few minutes." We rounded the cliffs and entered the mouth of an immense glacier-cut valley . . . and there lay our lake—frozen solid from end to end! Jim sharply banked the Beaver and headed out to sea toward the middle of Baffin Bay. "Surely you're not aiming for Greenland, Jim?" I said, my previous confidence in shreds.

"Don't worry about that, we'd never get halfway there," he replied. "But we had better find some open water soon, as the fuel gauge is reading almost empty. Take a look!"

After a few more anxious minutes over the sea ice, we did come across a wide lead, surprisingly free of icebergs or loose pieces of pack ice. Jim went straight for the water and landed not a moment too soon. So, finally, we sat gently rocking in the middle of an ocean of floating ice. The coastal mountains looked a very long way off—probably at least ten kilometres. It was almost still, only the lightest breeze rippling the water. Sitting there was a calming experience. Our hearts stopped racing and our pulses slowed back to normal.

Understandably, Jim wanted to refuel, get out of there, and fly back over land as soon as possible. He jumped out onto the pontoon, opened the rear cabin door, and began to wrestle with the first of the ten-gallon drums of avgas. Alarmed, I jumped out after him, got my hands on the climbing rope, and proposed to belay him onto one of the struts. It is necessary to explain at this point that the rear door of a Beaver is separated from the fuel cap by the wing struts. I saw that Jim, with a heavy drum of avgas on his shoulder, would have to duck beneath and around the struts, leaning out over the water.

"Wait a moment, Jim, and I'll tie you on for safety."

"No need," he replied, "I do this kind of thing all winter in the Gulf of St. Lawrence."

"This is different, Jim. If you slip off the pontoon, I will never be able to get you back on." But he insisted it was routine and told me to get back in my seat and leave him to it. I agonized while he repeated the trick three more times, to put forty gallons into the Beaver's main fuel tank, ducking beneath and around the struts with each heavy load. When it was over, he climbed in beside me.

He could sense my feeling of relief, then surprised me still further: "If I had fallen off, I'm sure you could fly the Beaver back to King Lake, right?" I replied that I was not so sure about that, and even if I could, I would have one hell of a task to explain how, and why, I had lost my pilot. That finally got him laughing. He started the engine and we began to race down the open lead, taking off with plenty of water to spare. He banked and headed inland to the coast, flew the full length of the frozen lake, and emerged over the icy waters of Dexterity Fiord. We then continued due south and climbed up above the inland plateau heading across Conn Lake toward the Barnes Ice Cap.

Once we were well away from the mountains and flying smoothly, Jim proceeded to give me tips on how to handle the plane if an emergency arose: "The takeoff would be easy. You've watched me and Spike do it many times. But when you approach King Lake, you should descend directly and smack down vigorously on the water on the first go. Don't worry about gently touching the water with the pontoons, nor about the landing conditions. Just get the damn thing down! If, in an extreme case, you break off the

pontoons, get out quickly and depend on your life jacket and hope somebody is close by with the canoe." This was all well and good, but I was relieved I never had to put it to the test.

We climbed slowly up the northeastern slope of the ice cap. As we crested the summit, we could see a most ominous bank of dark clouds stretching all the way across our western horizon. It would not be long before a devil of a storm would hit us, although we would have no trouble getting to camp well ahead of it. Distant lakes were glinting in the far west as we seemed to glide down the western slope of the ice cap. King River was soon immediately below us; within minutes, the Beaver's pontoons were cutting through the water and we quickly beached on the welcoming sand spit.

Spike was waiting on the dry margin and we went into the Parcoll for what was obviously to be a "council of war." Spike explained that he had just heard over the radio from Fox-2 that Hall Beach was already picking up snow and a major storm was on its way. It was very close to the date set as the end of the field season, so we opted for an emergency evacuation to Longstaff Bluff (Fox-2). Spike was worried that the storm would bring with it low temperatures and the risk of being frozen in on King Lake.

As there were only five of us plus the two aircraft, we were able to scramble and be off within forty-five minutes. The tents were left standing. All we took with us were the expensive instruments, notebooks, maps, and personal belongings. Fortunately, Spike had taken John and Pat and two loads of equipment to Longstaff while Jim and I had been cruising the northeast coast. George and Kent had also been retrieved from the Mary River camp and were on their way to Fox-Main, to connect with standard commercial flights to Montreal and Ottawa. But what about Mike and Bill, who were sitting on Flitaway Lake? Spike was insistent that we must leave them and sort out "that problem" later. He smiled grimly and said in his best philosophical manner, "Storms don't last forever. You and Jim can pull them out when the weather abates, but I'm dropping off my payload at Fox-2 and heading straight south for Frobisher, so let's get going!" So we did, landing at Fox-2 just as the first snowflakes arrived. Spike off-loaded, effected a very quick refuelling, shook hands, and departed with Roy for Frobisher.

To Flitaway Lake to "rescue" Mike and Bill

Jim and I had made the approximately 130-kilometre flight from King Lake to Fox-2 on Spike's tail without incident, buffeted by the very choppy air ahead of the ominous weather front. We had landed with sheets of spray hitting wings and engine cowling as the cloud base was approaching lake level. Snow had just begun to fall. Then we luxuriated in the warmth of the DEW Line station, with steaks, fresh bread, hot apple pie, followed by pool and table tennis, as well as hot showers and clean clothes. There was much forced laughter about Mike and Bill, still out there in the snow and wind, and some dark humour about how they would walk out—or perhaps overwinter! But the joking was symptomatic of a high degree of concern. I was certainly very anxious for them.

The storm raged for two whole days, during which time the surrounding landscape turned a glittering white. Our table tennis abilities improved remarkably and camaraderie with the station crew reached new heights. But what of Mike and Bill? No radio contact had been made. I had a telex from Ottawa asking that I return immediately (if only to struggle with an emergency Treasury Board meeting on which our 1964 budget would depend). I was in a quandary. John Andrews offered to go with Jim Cole to attempt a pickup of the two men, leaving me free to head for home. However, I had made a personal commitment at the start of our Baffin Island expeditions, in 1961, that I would never leave the field until I felt assured the entire party was en route for home, safe and sound.

On the morning of the third day, the sky began to clear. By noon it was a steely blue, without a cloud. But the wind was blowing and gusts of up to one hundred kilometres per hour were frequent. The Beaver could not take off in such conditions—some of the gusts were close to the plane's maximum speed! Jim's prognosis was that, when the wind did drop, the lakes would freeze quickly as the air temperature had remained very low. In that event, a floatplane would be of no use, and the task of retrieving Mike and Bill would take on a more serious hue.

The wind dropped steadily throughout the afternoon. The attempt would have to be made before nightfall because of the risk of an overnight freeze of the lake; at best, Mike and Bill would be starting to feel hungry. We set off about as late as we dared and made a harrowingly slow and bumpy flight northward, slow on account of the strong headwind. The familiar tundra landscape was now totally white, ghostly, hostile, and unrecognizable, except for the many lakes now showing white wave crests on the windswept water and allowing us to be certain of our direction. The land sloped imperceptibly up toward the ice cap; our tents looked stricken as we passed low over the base camp we had evacuated so quickly at King River. We crossed the Lewis Glacier and Flitaway Lake came into view. Clearly, we were headed for a very windy landing.

We flew low over the single tent, but all was still and silent. We hit the lake surface with a thud, bounced, and quickly lost forward motion in the teeth of the wind. We manoeuvred repeatedly to try to come close to the lakeside nearest the wildly flapping tent so I could have a chance of reaching dry land without getting totally soaked. Small pieces of ice were scattered across the water, seemingly jumping on the backs of the wave crests Worse, as a result of the recent drop in water level of the lake, large boulders poked above the surface all along the shore—a perilous situation. Furthermore, the spray was already freezing on impact with the pontoons and struts. The only thing keeping the lake open, in fact, was the wind. The air temperature was −3°C and had been below zero for some time.

To add to the difficulties, we backed into a large rock as we tried to manoeuvre for a landing as close to the shore as possible. This so ruined our twin water rudders at the back of the floats that Jim now had significant trouble steering on the water. He edged into the shore as closely as he dared and called out to me, "Sorry, Jack, but you are going to get a wetting! I only hope I can keep more or less on this spot so you can get back on again with the guys. But make it quick!"

I lowered myself off the pontoon, thigh deep in the icy water, and struggled the few paces to the shore, twice slipping in the lake mud and all but going under. I staggered onto the snow-covered shore and hobbled the twenty paces or so to the tent. As I struggled to pull the tent zipper down, a cheerful voice called out, "There's not much room but you're welcome. Come in and we'll do a brew of hot chocolate."

I replied, "Like hell! You've got two minutes to get out, and five minutes to collapse the tent and throw everything into the plane."

"Oh, a plane? We didn't hear it . . . thought you must have walked from King River, sorry!"

I took that to be a sample of clever Mike Church's warped sense of humour, though many years later he assured me it had been an innocent remark.

They emerged from the tent rapidly enough. We rolled up sleeping bags, tent, odd pieces of clothing, and fortunately waterproofed field notebooks into one big soggy bundle and tottered to the lakeside where Jim was managing to hold on. But he had been forced still farther out from the shore, and by the time I got my hands onto the pontoon, I was almost waist deep with spray breaking over my head. Bill and Mike heaved themselves up onto the pontoon, together with the tent and other effects. As they climbed into the rear, I struggled up over them and into the co-pilot's seat.

Jim, still struggling with the damaged water rudders, did some fancy manoeuvres with his wing flaps to clear the lakeshore and then, reducing throttle,

allowed the wind to drift us back to the end of the lake for maximum takeoff distance.

"Sorry, Jack . . . you'll have to go out again," Jim said. "Take your ice axe and knock off as much of the bloody ice as you can!" I tried that, but not for long as the spray seemed to be freezing on as fast as I was breaking it off. The "black frost"[5] encountered by my hometown Grimsby fishing trawlers north of Iceland in winter came to mind, as well as the dire prospect of slipping off the pontoon. Swimmer or non-swimmer, it wouldn't have made much difference.

Back in the right-hand seat I watched Jim open throttle, and away we roared. But looking sideways at the lake edge of the ice cap, we appeared to be hardly moving. We tensed, waiting for liftoff. Almost in slow motion, it seemed, the pontoons cleared the water. The plane seemed to just creep over pieces of floating ice and it was remarkable that we didn't hit any. We gained about fifty metres of elevation, then steeply banked and as we completed the turn; our groundspeed shot up from barely twenty-five to almost two hundred kilometres per hour, and we raced southward downwind.

I turned back to check on our two rescued researchers. They were sprawled out in a tangled mess of melting snow, soggy tent, and sleeping bags—both quite soaked, with ice in their hair. Mike called over the roar of the engine, "Dr. Ives, I would like to be the first student assistant to sign up for Baffin Island for next year." A warm glow spread through me. From Mike Church, that was praise indeed. In a way, though, this intrepid commitment was typical of almost the entire Baffin Island crew from 1961 until 1967.

All the while we were pursuing this hazardous adventure, everyone at Fox-2, regular DEW Line staff and branch members, had been anxiously awaiting our return. Apprehensions evaporated as we came in to land in the gathering dusk. Hot baths, steaks, and cognac for the four of us, then to bed, exhausted. All night long in my dreams I was seeing the Beaver hit a chunk of floating ice and plunge back into the lake. I asked Jim about it next morning.

"We very nearly did," he commented. "Another ten- or fifteen-kph windspeed and we would never have made it. But I am going south. . . . Thanks for the assist, Jack! Sorry to rush, but Longstaff Lake is freezing over. Flitaway will be solid by now, so I must get the plane back to Frobisher and home. See you next year!"

That was pure Jim Cole, one of the best Arctic floatplane pilots I have known. You could never tell when he was afraid, if he ever was. But he told me it was fear that kept his nerves taut and his mind keen, and fear had certainly been there on Flitaway Lake.

"Didn't you feel the same, Jack?" he asked.

"There was hardly time, Jim, but with me it hit afterwards."

Yes, there had been some very anxious moments. Only Mike and Bill appeared unperturbed—was that the folly, or the innocence, of youth?

The following morning, as arrangements were made for the entire group to catch a DEW Line lateral flight to Hall Beach, the officer on duty brought me a special-delivery confidential letter. Curiosity prompted me to open it immediately. I was shocked to the core. Dr. van Steenburgh had written to say that Dr. Nicholson had decided to take a year's sabbatical leave. Dr. Van wanted to see me as soon as I got back, indicating that he needed to discuss with me the prospect of my becoming acting director of the Geographical Branch.

So ended our Baffin Island field operations for the 1963 season. Once back at Hall Beach (Fox-Main) that morning, we caught the regular Nordair commercial flight south. We reached Ottawa very late on the same day, barely twenty-four hours after the nerve-testing rescue from Flitaway Lake and hardly a week after almost running out of gas over Baffin Bay. It seemed a little unreal to be back safe in a warm bed.

CHAPTER 6

❧ INITIATIVES AND GROWTH ❧
IN BAFFIN OPERATIONS, 1963–1964

Shortly after my return to Ottawa from Baffin Island in September 1963, I was invited by Dr. van Steenburgh for a personal consultation. I knew it would relate to his surprise message to me at Fox-2 about Norman Nicholson's sabbatical leave. He asked if I would be willing to serve as acting director of the branch until Norman's position became clear, although he sensed a strong possibility that the director's move to the University of Western Ontario would be permanent. He explained that accepting this position would limit my Baffin Island field activity and involve a considerable increase in responsibility, both national and international. He expressed regret for my youthfulness and lack of experience with upper levels of the civil service. The Geographical Branch, Dr. Van said, had been in an equivocal situation ever since its creation in 1947 and several sources within the government even opposed its continued existence. On the contrary, he believed that such an institution was of considerable value, and he would do all he could to support it, but he warned me that he would turn sixty-five—the compulsory retirement age—in less than three years. Assuming that Norman remained in academia, I would have limited time in which to build an institution strong enough to withstand the pressures for its dismemberment. In effect, my acceptance of his offer, he said, might leave me in an ambiguous position, perhaps for as long as a year. Yet, despite the considerable challenges I would experience, Dr. Van thought I was the most suitable candidate immediately available; assuming the decision would be a difficult one for me, he offered me a week to think it over.

I replied that I would give him an answer immediately if he could grant a single condition. This produced both a grin and a quizzical expression. I asked if I could interpret the term "acting" to mean that I was free to act. He laughed heartily and replied that I was already living up to his expectations. I thanked him for his confidence, shook his hand, and staggered back to my office in a daze. Once there, anxious that my uncharacteristically nervous reaction would be too conspicuous, I told Lil Murray, my secretary, that I felt unwell and would take the rest of the day off—a hard ploy to pull considering that I hadn't missed a single day since my initial appointment. And I certainly couldn't fail to notice her bemused expression as I left. I hurried home across the park at Dow's Lake and broke the news to Pauline.

I thus began the 1963–1964 winter in a quandary: How to continue the expansion of the Baffin Island project while at the same time accelerating my efforts to develop and strengthen the Geographical Branch as a whole? This was exacerbated, of course, by the question of how much of any "progress" could be retained in the event that Dr. Nicholson should return. Also, as acting director I would inherit a number of serious responsibilities. These included the positions of chair of the Canadian Permanent Committee on Geographical Names,[1] secretary of the National Committee of the International Geographical Union (IGU),[2] government representative to the Pan-American Institute of Geography and History (Organization of American States), and sundry civil service interdepartmental committees. Dr. van Steenburgh further insisted that I must represent the Geographical Branch at the International Geographical Congress, scheduled for August 1964 in London. This alone would probably rule out working on Baffin Island the next summer.

In addition to the international and interdepartmental responsibilities, I hoped to reorganize the publication of the *Geographical Bulletin,* the centrepiece of the branch's publication system, by converting it to a peer-reviewed quarterly journal. I wanted to work with Gerry Fremlin, the branch's senior expert on atlases, who was determined to reorient and rejuvenate the National Atlas program. Yet another concern was the creation of a national advisory committee on geographical research that ought to strengthen the extra-departmental relations of the branch, particularly with the universities. Ideally, it would have representatives from universities, government, and industry, with funding adequate to provide research grants for academic geographers. And during the alternate Monday morning meetings in Dr. van Steenburgh's office with the other five departmental branch directors, there was always the prospect of new initiatives and additional responsibilities.

This situation, together with the management of a staff now approaching a hundred persons, would require a delicate balancing act between Ottawa-based duties and Baffin Island fieldwork. Nevertheless, the winter proved not only challenging but enjoyable, although I felt a great sense of relief on April 16, 1964, when Dr. van Steenburgh informed me that Dr. Nicholson had submitted a written resignation and would remain on the faculty at the University of Western Ontario. He asked if I still would like a permanent appointment as director; if so, he would make a formal announcement forthwith. I did not hesitate. While my relationship with Norman had been a very complex one, and I had not received a word from him all winter, he had already made a significant contribution to the development of my career.

A bridge too far?

During the 1963–1964 winter, my main objectives were achieved or their subsequent realization assured. The transformation of the *Geographical Bulletin* was well underway. The legal and administrative work for the national advisory committee on geographical research was completed, and it was formally established by governmental Order in Council on April 14, 1965. And coincident with the use of a diagrammatic form of Champlain's astrolabe for the cover of the *Geographical Bulletin*, a large metal shield was designed and mounted in a prominent location high on the wall above Booth Street. (Fig. 21) This had required the express support of Dr. van Steenburgh. A new addition to the branch's atlas program was approved and work was already in progress. Reorganization of the Geographical Branch was well underway, and I was delighted to realize that most of the staff were enthusiastic.

This great feeling of progress, however, was somewhat offset by an invitation to a private meeting with Dr. James (Jim) Harrison, GSC director, immediately after receiving the news that Dr. Nicholson had resigned. Dr. Harrison observed that I had been presented with a great opportunity that could also solve the problem of overlap between the GSC Pleistocene Geology Section and my commitments to geomorphology and glaciology. He explained that he could arrange for the amalgamation of both groups to form a new "Division of Quaternary Geology" and that he had sounded out the senior geologists who would be involved and they would welcome me as

Fig. 21:
The Geographical
Branch shield,
constructed and, in
1963, mounted on
a wall high above
Booth Street. Note the
Champlain's astrolabe
design.

division chief.[3] Furthermore, because several of the Geographical Branch's economic geographers had resigned for academic appointments, the vacated positions could be used to recruit additional scientists for the proposed new geological division. He insisted that I was not a geographer but a geologist; what was more, the chance to become a division chief in the GSC was surely the opportunity of a lifetime, especially for someone in the relatively early stages of his career. He topped this off with a personal commitment to support my future plans for Baffin Island.

Dr. Harrison volunteered that he had discussed his proposition with Dr. Van, who had said he would leave the decision to me. Jim was paying me a considerable compliment and was very disappointed when I refused. I tried to explain that I could not regard myself as a geologist, and in any event, such a move was politically impossible for me as it would be regarded as a betrayal by all Canadian academic geographers. My rejoinder that the Pleistocene geologists were welcome to move into the Geographical Branch produced a friendly chuckle, with the admonition that *they* would never want to be called *geographers*. The game was tied. Despite this "merger and acquisition" attempt, and repeated ideological clashes in ensuing years, however, my relationship with Jim Harrison remained convivial throughout my time as director, and for decades after I left the civil service.[4] Nonetheless, this meeting did not bode well for the survival of a geographical entity within the federal bureaucracy. At the time, I convinced myself that the Geographical Branch would pull through, although I had not yet come to appreciate the full significance of Dr. Van's statement about his pending retirement. Meanwhile, because I considered that the confidence and enthusiasm of branch personnel were essential to the branch's continued existence, I decided to take on the additional psychological burden of keeping my worst fears to myself.

Preparations for the 1964 Baffin Island field program

By mid-winter, it was already time to prepare for the next summer's fieldwork in Baffin Island. The scale of operations was to be further expanded, and I accepted that I would not be able to go myself. An experienced field researcher was needed, therefore, to take on the role of expedition leader and perhaps, eventually, to take my place as chief of the Division of Physical Geography.

The outstanding choice for expedition leader was Dr. Olav H. Løken. He was then associate professor of geography at Queen's University, in Kingston. He had had experience in the Antarctic during the International Geophysical Year (IGY) (1957–1958) and we had worked well together during the 1958–1959 winter at the McGill Sub-Arctic Research Lab. Also, he had completed three seasons of excellent fieldwork in the Torngat Mountains, a place always close to my heart. I pinned my hopes on his finding the opportunity for research in Baffin Island a stimulating extension to his Labrador work. The first step was straightforward: would he consider one of the branch summer positions intended for university faculty, which would enable him to begin his own research in the eastern fiords of Baffin, as well as accepting overall leadership responsibility? His agreement was very welcome; after my confirmation as branch director in April, I was able to suggest that Olav consider replacing me as division chief on a permanent basis.

Concurrent with Olav's growing involvement, John Andrews had requested a year's leave so that he could begin his doctoral studies. The University of Nottingham (U.K.) and the supervision of Dr. Cuchlaine King was a natural choice.[5] I encouraged John to apply for a National Research Council scholarship, which he was awarded and which I was able to supplement with a year's leave from the branch on half salary. This gave him the equivalent of a full salary at a time when Canada's academic and research pay scales were higher than those in the United Kingdom. John's year at Nottingham would also strengthen the link with Cuchlaine and, given his three successful field seasons in Baffin Island, he would be able to complete his doctorate within a single academic year.

Representing the Geographical Branch at the IGU London conference

From April onwards, my pace of life accelerated. Brian Sagar, with six assistants, left for Fox-2 and the Barnes Ice Cap on May 10, 1964. Olav Løken set out in the same direction on June 15. Olav organized what became a major change of focus, emphasizing research in the mountains and fiords of the northeast, which would be supported by a new, well-equipped permanent base camp to be built at the head of Inugsuin Fiord. In the meantime, on Dr. Van's insistence, I made plans to represent the Geographical Branch at the International Geographical Congress in London. The very influential contacts that I made there proved a great help to my later career, although that was far beyond the reach of my imagination at the time (Ives, 2013).

In London, I was pleasantly surprised to be welcomed as an "up-and-coming" geographer by several senior British geographers whose names had been familiar from my undergraduate days, in the early 1950s. For instance, Lord Dudley Stamp introduced me to Dr. Alice Coleman, director of the Land Use Survey of Britain, and I was later able to invite her to spend a sabbatical year at the Geographical Branch in Ottawa. Another contact that turned out to be even more important was my reunion with Roger Barry, one of the 1957–1958 winter team at the McGill Lab, who was then teaching at Southampton University; consequently, he spent his 1966–1967 sabbatical leave as a guest of the Geographical Branch. Dudley Stamp also introduced me to Professor Carl Troll from Bonn, West Germany, the outgoing president of the IGU, and to Soviet academician Innokentiy

Gerasimov, the society's president-elect.[6] Gerasimov and Troll were later very influential in my election as chair of the IGU Commission on High-Altitude Geoecology.

Hilda Richardson, secretary-general of the British Glaciological Society (which in 1977 became the International Glaciological Society), organized a party for all attending glaciologists in the grounds of the London Zoo. Amid seemingly unlimited quantities of white wine and conviviality—as well as many curious four-footed, feathered, and web-footed friends— we talked well into the night and established many new liaisons. On reflection following my return to Ottawa, I came to understand what Dr. Van had been thinking when he insisted I go to the London congress. The benefits were many and continued to unfold throughout my entire career (Ives, 2013).

1964 Baffin Island summer fieldwork

In many ways, the 1964 season marked a major turning point in the Baffin Island project. Initially, due to a shortage of time, Olav Løken could be employed only in a summer faculty position. However, it was assumed that his new leadership appointment with the branch would become permanent, so his formal application documents were filed in Ottawa to be processed while he was in the North. Nevertheless, he assumed the key leadership role from the moment he left Ottawa. Although the overall aims of the Baffin project remained unchanged, the onset of the 1964 season saw its emphasis shift strongly toward the eastern mountains and fiords.

Olav selected an ideal site for a permanent base, at the head of Inugsuin Fiord. It was centrally located for operations within the fiords and along the outer coast, both north and south from Clyde Inlet, as well as for the Barnes Ice Cap. Fox-3 (Mid-Baffin, or Dewar Lakes), rather than Fox-2, became the closest DEW Line station to Inugsuin Fiord and thus now served as our main logistical hub. It also provided

good access, along with Fox-2, to areas farther west that would come under study in future years. By this time, the Baffin Island project seemed to be sufficiently well entrenched that a "permanent" base camp at the head of Inugsuin Fiord was eminently reasonable.

The designation of the years 1965–1975 as the International Hydrological Decade (IHD), another United Nations (UNESCO) attempt to augment international scientific cooperation, provided an impetus for glaciological studies. It justified another increase in both our budget and our staff, necessary if Canada was to play an effective role in the glaciological aspects of hydrology. Dr. Van supported an enlarged research program on the assumption that snow and ice were, without doubt, essential elements of the hydrological cycle, especially in a northern country such as Canada. The additional resources facilitated expansion of the work on the Barnes Ice Cap and the Lewis Glacier as well as preparations to expand the glaciological program into the eastern mountains and fiords.

Olav consequently began to plan a glaciological transect across the eastern Canadian Arctic at middle-high latitudes through the heart of our Baffin field area; this would complement Fritz Müller's work on Axel Heiberg Island, now in its fifth year, as well as the Arctic Institute's work on Devon Island and in the Yukon. On Gunnar Østrem's urging, Olav and I were also encouraged to argue that any national program must include a transect across Canada's western mountains: the Rockies and Coast Ranges from Alberta to Vancouver Island. There should also be a firm commitment to produce a national glacier inventory. These additional components to the Baffin Island undertakings became important elements of Gunnar Østrem's responsibilities as head of the Glaciology Section of the Division of Physical Geography the following year.

Brian Sagar's ice cap group left Ottawa in mid-May by commercial flight and flew from Sept-Îles by chartered DC-3 on ski-wheels direct to Fox-2. The DC-3 was retained to deliver supplies to the site of the new Inugsuin base camp as well as to the field parties at the north end of the Barnes Ice Cap. A Beaver on ski-wheels reached Fox-2 on May 24, the day before the DC-3's charter ended and it was due to head south.

This was the second season that Chris Bridge was Brian's assistant on the ice cap. Chris was to play a prominent role later in the summer. Dave Harrison, Mike Church, and Bill Rannie, by now "old hands," reached Fox-2 on May 28. Jim Peterson, Martin Barnett, and Tim Fielding—graduates of the McGill Sub-Arctic Research Lab who had obtained their first wilderness experience in Labrador-Ungava, with Ed Smith and Bruce Smithson—arrived on June 10, along with newcomers D. B. Fish and B. Trnavskis. Olav, with Pat Webber and his botanist assistant, John Morse, arrived on June 15. The remaining assistants included Angus Cherrington, D. J. Perraton, and Peter Hall, bringing the total to eighteen, plus the aircrew.

Once again, the scale of operations had grown and, although the planning for complicated field logistics had been fine-tuned over previous years, a remarkably cool season somewhat crimped the anticipated scale of the results. Nevertheless, significant advances were made in terms of linking sections of the Cockburn Moraine formation to isostatic changes in sea level during the later phases of the "last" ice age. Olav was enthusiastic about the potential for future work in the fiords and along the outer Baffin Bay coast.

Toward a permanent base camp at Inugsuin Fiord

Construction of the new base camp at the head of Inugsuin Fiord got underway in late May. Two large double-walled tents were erected with wooden frames. A small A-frame hut, prefabricated in Ottawa to Gunnar's specifications, consisted of tongue-in-groove boards and was easily put together in the field.

The department's equipment store managed to escape Gunnar's eagle eye, however, mysteriously producing a structure that was much higher than the design he had drawn up. While it provided welcome extra storage space under the unexpectedly high peaked roof, heavy guy-wires were needed to counteract the structure's increased wind resistance. To ensure extreme tent anchorage, the Atlas Copco portable pneumatic drill was used to drill holes through large boulders, for grounding the guys. Two boats were added to the growing mass of equipment: a six-metre aluminum vessel and a four-metre rubber boat, both with 10-horsepower outboard motors. The boats were also equipped with oars and were of great value for crossing the fiord once the ice had dispersed sufficiently.

This year, Brian moved from his usual base on the crest of the Barnes Ice Cap to its northern margin. Here he was within easy walking distance of the Flitaway Lake and Lewis River camps where Gunnar was instructing the student assistants to measure the discharge of the very turbulent Lewis River in addition to setting up an expanded stake network on the glacier with which to record the rate of ice melt. Brian's group, whenever they had time to spare, provided additional help for what was a very labour-intensive undertaking. Brian also extended the micro-meteorological network from the ice cap up the slope of the nearby hill. And once the meteorological instruments were functioning, he arranged for daily radio transfer of weather data to Frobisher via the Clyde River station, a practice that was much appreciated by the Department of Transport and the DEW Line. Chris Bridge, now well trained and familiar with the "met" instrumentation, took full responsibility for the ice cap work for the rest of the season, giving Brian a chance to return to Ottawa for the joyful event of his marriage to Norma.

Before the field parties left Ottawa for Baffin Island, it had been suggested that an attempt should be made to record some of the operations on colour movie film. To this purpose, when it was discovered that one of the students recruited as a Baffin field assistant, Angus Cherrington, was a semi-professional photographer with some experience with cine-photography, the plan was put into effect. A Bell and Howell 16mm cine-camera was purchased and Angus was asked to shoot some footage of the expedition insofar as his glaciological commitments permitted. Once the film had been developed and we had screened the results in Ottawa the following fall, it was seen to be so impressive that I regretted Angus had not been appointed solely as expedition movie photographer. There are always lessons to be learned.

A small glacier situated on the opposite side of Inugsuin Fiord from base camp had been identified from air photographs as a possibility for intensive study as part of our planned contribution to the IHD program. Olav made a field inspection early in the season and confirmed that it was a good choice; it was later officially named Decade Glacier. Ablation stakes were installed up to a height of 888 metres asl and, for the period from August 1 to 24, net ablation ranging from 325 millimetres to 40 millimetres was recorded along the vertical profile. Bill Rannie was flown over from the Lewis River camp to reconnoitre the glacier's meltwater stream for possible hydrological studies and site selection for automatic gauging stations. A large number of glacier photographs were taken throughout the nearer fiords and mountains, to be added to the branch's glacier inventory. Olav made a special flight to the area around Sam Ford Fiord and Walker Arm to extend this undertaking.

Interpretation of the relationships between late ice age glacier retreat and changing sea level was greatly enhanced by Olav's identification of a large complex beach deposit at the head of Inugsuin Fiord. He studied the stratigraphy through almost one thousand metres of silt, sand, gravel, and pebble layers, tracing many of the specific strata to their respective former sea levels, from which seashells were collected for radiocarbon dating (Løken, 1965). This was the first systematic study of glacier moraine/sea level relationships in the eastern Canadian Arctic,

and it provided a vital base for more extensive studies in following years.

Plant ecology and vegetation benchmarking

Pat Webber, assisted by John Morse, completed his second season with the Geographical Branch operation. His work had already shown the value of both interdisciplinary studies and government-university collaboration. It had deepened our understanding of lichenometry and resulted in publication, with John Andrews, of several academic papers on this topic as well as the application and refinement of Roland Beschel's pioneering studies. Pat had also added to the knowledge of the plant ecology and plant geography of the region, which became the topic of his doctoral dissertation at Queen's University some time later. In particular, he had greatly expanded the known range of many plant species, including *Carex gynocrates*, *C. glacialis*, *Woodsia alpina*, and *Puccinella andersonii*. Some of these were found more than two hundred and as much as seven hundred kilometres beyond their previously known limits (see chapter 11).

One of the most remarkable of Pat's achievements, for the time, was the somewhat inadvertent establishment of a huge permanent vegetation benchmark for longterm determination of the effects of climate change on plant growth and species composition. He had set up almost a hundred permanently marked vegetation quadrats at varying distances from the 1963 northwestern margins of the Barnes Ice Cap, from which he archived an extensive data bank. It was inadvertent because Pat was pursuing hypotheses about the nature of the Arctic plant community *per se* rather than about plant succession on ageing surfaces or under a changing climate regime. It proved possible to revisit most of the quadrat sites and repeat the original observations up to, and even beyond, fifty years later. To this were added summer climate data. By the time of the Fourth International

Polar Year (2007–2009), it proved possible to relate an increase in number of plant species, percentage of ground cover, and enhanced plant growth to the influence of a warming climate in combination with lapsed time since the various sites were exposed following retreat of the margins of the ice cap. This important result is discussed in greater detail in the concluding chapter of this book (P. J. Webber, personal communication, October 2010).

The 1964 season proved to be very successful, despite the unusually cool weather, which restricted aircraft operations. There was one potentially serious incident: once again a Beaver on pontoons hit a floating piece of ice while taking off from Flitaway Lake, on August 10. The pontoons were so badly punctured that the plane sank on landing at Fox-2. However, no one was injured, and the readily accessible DEW Line heavy equipment ensured that the Beaver was back in operation within two days.

This incident greatly reinforced the early awareness of the need for helicopters. On their return to Ottawa, Olav and Brian raised with me the question of whether the approaching IHD program would justify the necessary increase in budget. I decided that the occasion was ripe for once again approaching Dr. Van. This proved successful, and a large increase in budget was approved to allow for helicopter operations to coincide with the start of the IHD program the following year.

During the early summer of 1964, an entire building—fifteen metres by five metres—for the Inugsuin base camp was prefabricated in Ottawa in time for it to be loaded onto the government icebreaker *CCGS John A. Macdonald* in Montreal. It consisted of three rooms: a dining room with fully equipped kitchen, a laboratory-library–reading room, and a sleeping room with six double bunks. Very firmly constructed, it was wired for electricity to be provided by a field generator. The *John A.* was able to sail up the entire length of Inugsuin Fiord during early September, after resupplying the small settlement of Clyde River. All the material was unloaded within a few paces of the pre-selected construction site. There were also two more prefabricated A-frame huts, one destined to be lifted by helicopter the following summer, together with freeze-dried food and aviation fuel, to a point high on the Decade Glacier. These buildings and helicopter support would provide an unprecedented facility upon which to substantially enlarge our field operations. The main building at the head of the fiord stood up to the severe weather of fall and winter for many years and was used by Inuit spring hunting parties from Clyde River for decades after it had fulfilled its primary function.

Winning the struggle for gender equality in the Arctic

As I witnessed the incorporation of more than a dozen university undergraduates and graduates into Geographical Branch summer work in the Arctic, I became increasingly concerned that women students were still entirely prohibited. This contrasted with my earlier experience in a university environment. Women had worked out as well as men in isolated field situations in Labrador-Ungava, though they all had been wives (Ives, 2010). Perhaps more relevant, women had made contributions fully equal to those of men with me on the University of Nottingham expeditions to southeast Iceland in 1953 and 1954.

I broached the topic with Bob Code, departmental director of personnel, and Jim Harrison regarding the GSC's policy, questioning the reasons for denying field experience to women. Jim expressed strong opposition, explaining that the GSC would never allow women on field surveys. His main point was that it would pose "family-related" problems. Women whose geologist husbands were frequently away from home, for three months or more at a time, were single-handedly running the household and taking care of children; some wives, at least, would not tolerate mixed field parties—so went the argument.

The inclusion of women geology students would also greatly complicate logistics.

Bob Code took a different tack. He first insisted that women simply were not physically strong enough to operate effectively under rugged field conditions, often in extreme isolation. I countered this from personal experience: Pauline, my wife, had been with me for three long summers in central and northeastern Labrador-Ungava, as had Inger Marie Løken with Olav, and my undergraduate tutor Dr. Cuchlaine King, along with two Cambridge University undergraduate women, had squarely matched the men in performance in Iceland under physically demanding conditions. Next, Bob put forward those old fallback arguments: field toilet facilities and our dependency on overnight stays at DEW Line stations would place any woman at risk of harassment, or at least embarrassment. I again countered with my 1958 experience at the isolated Department of Transport (DOT) weather station on Indian House Lake in central Labrador-Ungava. Up there, the five permanent male staff, who were alone for up to six months at a time, put on white shirts and ties and decorated their dinner table with white linen in response to an unplanned visit of a mixed group that I had arranged from Schefferville.

Eventually, I related that Cuchlaine King had achieved international recognition for her publications in geomorphology and glaciology, including major textbooks, and that she had the experience to make an important contribution to the Baffin Island project. At this point, it was finally agreed that an exception could be made in her case. Adroitly seizing the advantage, I followed up by insisting that it would be totally "improper" if she did not have two women students as tent companions and field assistants.

And so the barrier was breached, initially as an exception for Cuchlaine. From then on, recruitment of women for summer field party positions became an annual event, but only within the Geographical Branch.[7] Cuchlaine arrived in Ottawa in mid-May

1965 and with her usual enthusiasm began to prepare for flying north in late June.

The struggle with the higher powers I have just related, however, had been made the more difficult because of an earlier tragic event involving the loss of two women graduate students in the Arctic. In 1960, Anne-Marie Krüger, a McGill graduate student and colleague of mine, was trying to arrange for independent fieldwork along the northern shore of Great Bear Lake, in the Northwest Territories. To augment her limited financial resources based on an AINA grant, her research supervisor, Professor J. Brian Bird, and Mike Marsden, director of the AINA Montreal office, had been able to persuade Dr. Nicholson to provide additional support. This was accomplished by seconding one of the graduate women who had been recruited for summer employment at the branch, Joan Goodfellow, to accompany Anne-Marie as field assistant. This arrangement had also justified provision of federal financial support, although Joan's participation was classified as fieldwork undertaken through the Arctic Institute rather than the Geographical Branch. Both drowned in a canoe accident. Joan's father was a Member of Parliament and the episode had even been raised on the Hill, where there was general opposition to the notion of women working in the Arctic.[8]

Branch expansion from an unexpected source

The 1964–1965 winter also saw the creation of a new toponymy (place name) division within the branch, composed of several geographers from the regular staff and the former Surveys and Mapping Branch toponymy support staff. The latter's responsibility had been to provide assistance to the Canadian Permanent Committee on Geographical Names. Since the establishment of the Geographical Branch in 1947, its director had served as chair of this national naming committee, whose members were appointed by each of the provinces, the Yukon and Northwest Territories (as a prerogative of the federal Department of Northern Affairs and Natural Resources), and several other federal units, including Surveys and Mapping and the Geological Survey. The new arrangement appeared as a useful rationalization, and Dr. J. Keith Fraser of the Geographical Branch became chief of the toponymy division. It added a significant place-name research arm. Considering the size of Canada, and especially its vast Arctic and Subarctic territories, processing the hundreds of new names that were submitted every year was heavy work. Each had to be scrutinized for correct derivation and its accordance with the quite formidable "rules of nomenclature." For instance, names of living persons were strictly proscribed, with very rare exceptions.[9]

Yet another step forward: A peer-reviewed quarterly publication

Another of my initiatives, as mentioned earlier, was the conversion of the *Geographical Bulletin* from its old format as an in-house occasional publication to a peer-reviewed quarterly. This involved a total redesign—including a new cover featuring Champlain's astrolabe[10]—as well as setting up the international editorial review board and producing four issues a year tied to a strict schedule. The senior departmental editor, friend, advisor, and cynic Doug Shenstone, assured me that the proposed production schedule

would never be met within the chaotic federal government bureaucracy. He was very nearly correct, as the third issue of 1965, our first year, was perforce numbered "3 and 4."

I certainly did not regard these accumulating responsibilities as unfortunate distractions from the main objective: Baffin Island. For these and the many other tasks of branch director, a certain high degree of stamina was required. But there were also fascinating distractions—some, of course, being merely coffee-break accounts. One of my favourites was a toponymic conundrum. A time-honoured and somewhat sarcastic place name proposed by a Royal Navy captain during the golden age of the Empire's penetration of the Canadian North could not be fitted onto a modern map. Should the very verbose name be "edited" or eliminated? It was, after all, "The Sons of the Clergy of the Episcopal Church of Scotland Islands."[11]

And so the 1965 Baffin Island field season approached. It seemed we were at the crest of a wave: we had the powerful support of Dr. Van, a rapidly expanding Glaciology Section, a quarterly journal, prospects for a politically important national advisory committee on geographical research, the stirrings of enthusiasm throughout the organization, a breakthrough for women in Arctic fieldwork, growing international and national recognition, and above all, a greatly expanded Baffin Island operation—interdisciplinary, international in tone, and finally anticipating full helicopter support.

Fig. 22: Inugsuin Base Camp. The ideal site for a base camp, at the head of Inugsuin Fiord. One of Gunnar Østrem's A-frame huts sits to right of the white tents. This inner section of the fiord, far from the outer coast, provided a Baffin-type "Mediterranean" climate. (Photo: July 1966)

CHAPTER 7

❧ GLACIOLOGY IN THE ROCKIES ❧ ADDED TO BAFFIN STUDIES, 1965

The first year of the IHD opened as plans for the Geographical Branch's new venture into the Canadian Rockies and Coast Ranges were being refined. Gunnar Østrem arrived in Ottawa from Stockholm on New Year's Day, 1965, this time with his entire family. I had found my way through the Civil Service Commission's labyrinthine channels for specialized appointments and, as part of Gunnar's special pleading for Swedish and Norwegian assistants who had previously worked with him in northern Scandinavia, six of them were added to our glaciological field team, including Stig Jonsson, Wibjörn Karlén, and Randi Pytte. Randi was part of the "women in the field" breakthrough. The strong Scandinavian personnel contingent was vital to what would be an exacting undertaking that we hoped would produce almost immediate results. Keith Arnold, another member of the permanent branch staff, who was temporarily distracted from his Polar Continental Shelf Program love affair with the small Meighen Island ice cap in the High Arctic, provided additional, well-seasoned support. Kent Sedgewick, Tijs Bellaar Spruyt, Robert Gilbert, and Chris Cambray were recruited from the now rapidly growing number of students from Canadian universities applying for summer employment.

Gunnar's objective was to select representative glaciers for detailed and long-term mass balance studies that would provide a cross-section from the eastern rain-shadow side of the Rockies with a continental climate to the wet maritime western Coast Ranges. Factors that would affect the choice of individual glaciers during the field reconnaissance included distinctiveness of their accumulation areas, size, nature of their meltwater streams, relatively smooth surfaces for safety and ease of foot or ski travel, availability of early records such as photographs, and accessibility for approach on the ground. It was hoped that six suitable glaciers could be located. This would require reconnaissance inspection of a large number of glaciers across some of Canada's most rugged terrain.

Glaciological fieldwork in the Rockies and Coast Ranges

The work began in Ottawa by examining all of the available 1:63 360 topographic maps: that is, one inch equals one mile (the adoption of metric by the federal government came later). Some areas of the cross-section had no adequate map coverage and so had to be eliminated. Nevertheless, some three hundred glaciers were selected for more intensive review. This led us to the National Air Photo Library. The air photos clearly demonstrated that many of the three hundred were unsuitable because of severe crevassing, a feature not shown on the topographic maps. In the end, thirty individual glaciers were identified for final examination on the ground. At this stage, the operation shifted to the Rocky Mountains, with Gunnar taking with him as an assistant Stig Jonsson, an experienced skier who had worked with him in Arctic Sweden.

In mid-April, I received an urgent telephone call from Gunnar, who was then at Lake Louise. After his initial vivid and enthusiastic account of the entire proposed transect of Canada's mountainous west, he explained that he was facing problems of accessibility. He had realized that an intensive reconnaissance survey would involve field visits to each of the provisionally listed glaciers before he could proceed with confidence to the final selection. It had become apparent that this would require extensive helicopter support. He needed to find out if his budget could be increased, in order to enable him to set up a series of local helicopter charters. I said I understood the necessity and that he ought to go ahead.

"But how many hours of helicopter charter time can I plan to use?"

"As many as you need!" I rejoined.

And he needed a lot—too many for either of us to remember now! I told him "not to worry," as I would go cap in hand to head office, and that I felt confident. To this day, Gunnar still talks about my response and how it totally astounded him, as he had fully expected questioning, compromise, or apologetic regret.

As a result, five glaciers were firmly identified: from east to west, Ram Glacier, northeast of Lake Louise (one of the most easterly); Peyto Glacier, some thirty kilometres north of Lake Louise; Woolsey Glacier, in Mount Revelstoke National Park; Place Glacier, northeast of Pemberton; and Sentinel Glacier, between Pemberton and Vancouver.

Once the final selection had been made, the rest of the team joined Gunnar and Stig, and all was set for an energetic summer. Trigonometric stations were established and precisely surveyed for future special air photography and preparation of contoured maps; numerous snow pits were dug; accumulation and ablation were measured continuously on two glaciers and intermittently on another; and meteorological data were collected from selected locations. Equipment and many specialized instruments had been brought from Stockholm, and given the intensive undertaking that extended across the entire width of the western mountains, the tightly knit and experienced field team more than justified itself. Valuable experience was also transferred to the Canadian summer student assistants.

One special outcome was that the detailed survey of Place Glacier, along with air photos taken later in the summer, became the basis for the first metric map ever produced by the Surveys and Mapping Branch. It was a challenge to negotiate this switch to the metric system with the very conservative federal surveyors and mappers, but I doubt that anyone could have been more persuasive than Gunnar Østrem. Much assistance for the fieldwork was provided by the Canadian National Parks Service and the DOT, who supplied all the standard meteorological instruments. Specialized equipment was borrowed from the Norwegian Water Resources and Electricity Board. The western field research that year was an outstanding example of interagency and international collaboration.

Olav Løken joined Gunnar at the end of August for an inspection of the ongoing work on the Peyto Glacier. This led to a short journey across the Canada–United States border to meet with Dr. Mark Meier, a leading American glaciologist. The ensuing negotiations led to adoption of internationally uniform field methods so that comparative assessments could be made. The stage was set for a major expansion in the 1966 season. Gunnar had also spent most of May on Baffin Island. The scientific significance of the glaciological efforts of the 1965 summer, both in the western mountains and in Baffin Island, is described in more detail in chapter 11.

Baffin Island 1965 field season

The 1965 Baffin Island expedition was the single largest operation organized by the Geographical Branch to that date. Seventeen student assistants were recruited from eleven Canadian universities, several of them having been with us on previous Baffin operations. Of the permanent staff, Olav Løken took over the position of expedition leader and prime organizer; John Andrews had returned the previous fall with a doctorate from the University of Nottingham based on his earlier Baffin research and was more than ready to make up for missing out on Baffin in 1964; Brian Sagar reoccupied his now regular Barnes Ice Cap position; Gunnar pushed ahead with the new IHD components; Martin Barnett returned with Dave Harrison, now on permanent staff; while George Falconer managed supplies and logistics with great skill. Dr. Cuchlaine King arrived in Ottawa from the UK in mid-May and by June 26 was setting up her first field camp at Eqe Bay, west of the Barnes Ice Cap, with two assistants: Jane Philpot (later Buckley), newly recruited to the permanent staff, and a University of Toronto undergraduate, Wendy Jocelyn (later Smith).

Once again we were able to provide field support for graduate students who undertook their own individual research: botanist Robert Hainault, of the University of Ottawa, and geologist Norman Gray, from McGill. They worked on independent thesis projects while relying on our logistics and field facilities. There were two other innovations, as mentioned previously: construction early in the season of the splendid base camp building at the head of Inugsuin Fiord and our first dedicated helicopter charter. A professional carpenter was flown in to supervise construction of the base hut at Inugsuin. After four days, the building was sufficiently advanced that he could leave its completion in the hands of the regular expedition members.

As in previous years, a chartered DC-3 on ski-wheels was used to transfer personnel and heavy equipment, especially to the new Inugsuin Fiord base camp and to the ice cap. It was backed up by a Cessna 185, also on ski-wheels. During the open-water season, a Beaver (CF-GBF) from Northern Wings, in Sept-Îles, provided long-range reconnaissance, camp moves, and, later, assistance with the evacuation of the widely dispersed field parties toward the end of August. It was piloted by my old friend from Baffin '63, Jim Cole.

George Falconer, Gunnar Østrem, and Brian Sagar, together with seven assistants, had reached Fox-2 by May 12. Dave Harrison left Ottawa by chartered Cessna 185 on ski-wheels on May 17, reaching Fox-2 a week later. Michael Church, flying the commercial route from Ottawa to Montreal to Hall Beach, arrived on May 18, so that an early start to the fieldwork was assured. Gunnar and Chris Bridge had also reached Inugsuin and the Decade Glacier. Heavy work preparing food and equipment caches at Fox-2 continued under George's direction. The DC-3 was able to return south on May 22, while the Cessna arrived at Fox-3 the following day. Gunnar departed on May 29 for Ottawa and to oversee the new field program in Alberta and British Columbia. During the hectic period from May 22 until June 7, nine more personnel arrived. Olav reached Inugsuin base camp on June 8 and took control over the entire complex of operations. The 1965 season comprised a complicated series of separate activities, and only a few of the main ones are detailed here. However, credit must be given to Olav's organizational skills, backed up by George's indispensable management of the logistics.

Brian Sagar reached the north end of the Barnes Ice Cap on May 19, together with M. Birtles, Mike Church, Barry Goodison, Peter Lewis, and G. Moroz. Mike, Barry, and Peter quickly reoccupied the Lewis River camp and began a more extensive glacio-hydrological study. Brian and his two remaining assistants tackled the ice cap. This involved further extension of the accumulation and ablation measurements of the previous years, often hindered by inclement weather and poor visibility. Nevertheless, the now very large

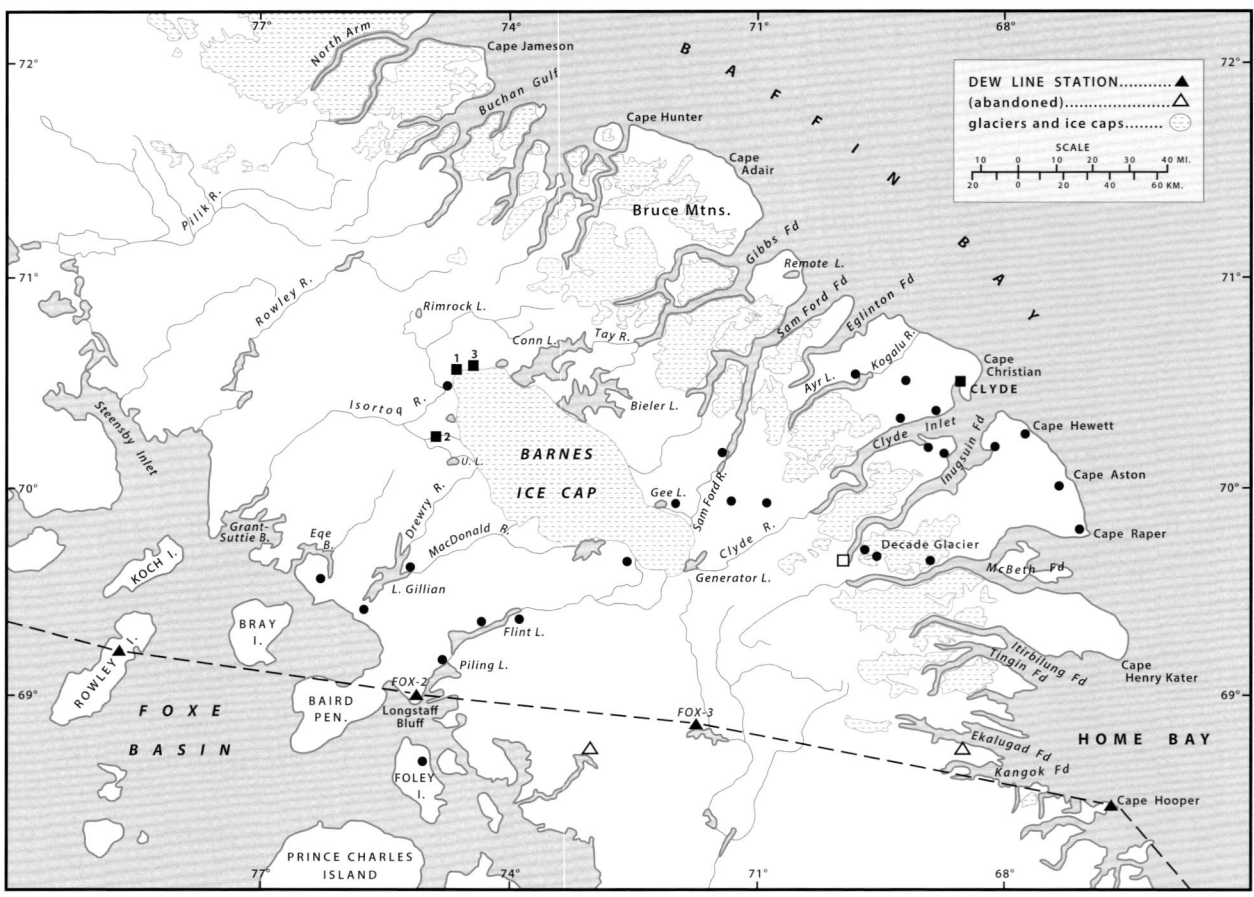

MAP 7: Greatly expanded activities from 1965 to 1967 (see also Map 8). The black squares represent (1) Flitaway Lake, (2) King River, and (3) northern margin of Barnes Ice Cap (glaciology); the open square shows location of Inugsuin Fiord base camp. (See photo facing page 99)

array of stakes was surveyed; 1,600 snow-depth and 400 snow-density measurements were recorded, and samples were collected for tritium dating. (Map 7)

On June 10, all the ice cap personnel were flown to Gee Lake, on the northeast edge of the south dome, where they made the preliminary moves to establish a northeast-to-southwest transect across the entire width of the ice cap. On June 14, Olav relieved Brian, who returned to Ottawa via Fox-2. From this point, Olav completed the transect, inserting thirty-six equally spaced stakes. This enabled him to confirm a previously introduced hypothesis that the mass balance of the ice cap was strongly asymmetrical, with a positive balance northeast of the crest line and

a tendency to a negative balance to the southwest. This fitted with the geomorphological fieldwork that indicated a progressive move of the region of greatest ice thickness from the middle of Foxe Basin toward the mountain rim of northeastern Baffin during the final 15,000 years or so of the last ice age and that this process was continuing into the mid-1960s. The southern traverse of the ice cap was completed on June 18, again in very difficult weather.

Cuchlaine and John and their assistants focused on the area west and south of the Barnes Ice Cap and along the Foxe Basin coast. Their combined work involved a series of camps between Eqe Bay in the north and Piling Lake, Foley Island, and the Tweedsmuir

Islands in the south. Extending the earlier work in the Steensby Inlet area (1961) and Grant-Suttie Bay (1963), it provided a well-documented account of late-glacial entry of the sea into Foxe Basin, its overlap onto southwestern Baffin Island, and the contemporary retreat of the Baffin Island remnant ice sheet that was eventually reduced to the Barnes Ice Cap of today. It facilitated the construction of precisely determined isobases and proof that lobes of the ice sheet remained in contact with the sea between 7,000 and 5,500 years ago. Configuration of these isobases, which run approximately parallel to the general alignment of the present southwest coast, confirmed the original hypothesis that Foxe Basin had served as a centre of outflow of continental ice during the maximum of the last ice age.

Meanwhile, the Spartan Air Services helicopter (Bell G2A) had been flown disassembled from Montreal to Frobisher inside a commercial freight plane. Pilot David Harrison[1] with engineer Tom Murray reassembled and test flew the machine at Frobisher, making a successful positioning flight northward up the western coast of Baffin Island to Fox-2 and then eastward to reach the Inugsuin base camp by June 28. The very first day of operations at Inugsuin, taking full advantage of fine weather and twenty-four-hour daylight, David logged a full eight hours of time in the air with several of the field parties. This was an encouraging start to the helicopter support work, though it wasn't long before there was another airborne adventure story to be related.

Olav had initiated studies on the sedimentary succession exposed along the outer coast between Cape Christian and Cape Roper. On July 1, Olav and MacHattie were flown in David Harrison's helicopter to the coastal lowlands north of Clyde Inlet. Olav's objective was to investigate the low-lying sediments of the outer coast first identified by Professor Dick Goldthwait during the 1950 AINA expedition (R. P. Goldthwait, personal communication, January 5. 1961). As recalled by David some years later:

This work on the east coast has vivid memories for me. July 2 was a long foggy day of fieldwork, and it was still only my first week in this part of the North. While Olav and his assistant picked antique seashells from the cliffs, helicopter FCK and I had mainly sat around on a beach enveloped in grey mist. Came the end of the day, the three of us got back in the helicopter—tired, hungry, and ready to fly back to a warm camp. The engine, by now quite soaked in foggy droplets, had other ideas: it failed to start. Three or four more attempts and the battery was dead flat—and with it the radio. We could be waiting a long, long time for a pickup. Olav stoically but politely dismissed my (not very bright) idea about hiking along the shore to Cape Christian (the U.S. Coast Guard LORAN base many miles south), knowing there were several rivers and outlets with open leads that would be very unsafe to cross at this time of year. He just pointed 90 degrees out to the vast expanse of sea ice and announced, "No! we'll walk out there and continue on the sea ice—it will be crunchy but we shouldn't fall through." So we crunched along for hours and hours. It was around midnight the following day when we walked, pretty fatigued, into the overheated mess hall and astounded the guys on the base who were watching the latest movies and supping on beer and pretzels. It was the morning of the Fourth of July. I am sure we had been stalked by a polar bear for much of that walk. Fortunately our number was not up. Olav, of course, seemed to take it all totally in his stride. (Personal communication, March 12, 2011)

Luckily, another helicopter (with a spare battery) was soon diverted from some other northern operation so David could pick up his stranded machine and get back to Inugsuin with Olav and MacHattie; very little time was lost, and David had learned two valuable lessons in the geography and climate of the High Arctic. There was never another flat battery (or forced march) in the three years of his work with us.

Coastal work with marine fossil dating

Olav's east coast investigation resulted in examination of the complex wave-cut section of sea cliffs stretching for more than thirty kilometres northwest from Cape Christian. The freshly exposed cliffs, regularly undercut at high tide, displayed a great variety of deposits ranging from bouldery glacial clay and coarse gravel beds to sand and clayey silt layers. The stratification was distinct and nearly horizontal and could be traced for considerable distances. Complete marine fossils and shell fragments could be found in most of the individual layers. Exposures of the bouldery clay were easily accessible along the lower section of the cliffs and yielded many marine fossils (shells). Higher up, a thick section of stratified sand contained large numbers of such fossils, many upright and, therefore, in apparently living positions. Olav made large collections that were identified as cold, or arctic, species. Only four different species were present: *Hiatella arctica*, *Mya truncata*, *Serripes groenlandicus*, and *Astarte borealis*.

The coastal work was extended south of Cape Christian on the far side of Clyde Inlet for nearly eighty kilometres, to Cape Aston. Here, large deltas graded into what was interpreted as a higher sea level, about sixty-five metres above present sea level. The delta strata yielded similar assemblages of marine fossils. Following the field season, fossil materials from both Cape Aston and the Cape Christian cliffs were dated at 50,000 years and 54,000 years before present (BP), respectively.[2] The Cape Christian and Cape

Aston results, the first of their kind from the Canadian North, provided a controversial preliminary supposition that parts of the outer coast, between the fiord mouths, had not been covered by glacier ice for at least 50,000 years. This result complemented my own efforts on the mountain peaks and uplands. Nevertheless, later work has led to claims that these conclusions need to be modified (see chapter 11).

Additional work on the outer coast and within the fiords by both Olav and Ed Smith (under Olav's supervision) provided much more detail on the interrelations of land, sea, and glacier ice during the closing period of the last ice age. During this period of about 18,000 to 7,500 years ago, the main outlet glaciers continued to penetrate into Baffin Bay while the land between the fiords was free of ice. This was indicated by the late-glacial marine limit: as high as eighty to eighty-five metres above its present level away from the fiords and only about twenty-five metres (or less) within the fiords. This implied that the land was rising in relation to the sea while the glacier tongues still extended to and beyond the fiord mouths, thus preventing access into the fiords by the rising sea. These initial findings prompted an invitation to Dr. Rolf Feyling-Hanssen, an internationally recognized expert on marine micro-palaeontology from Aarhus, Denmark, to join us the following summer for an intensive study of the coastal deposits.

The 1965 summer was also the first time in Arctic Canada that measurements of slope development (talus slopes) were initiated, this time by Martin Barnett in the vicinity of the Inugsuin base camp, with provision for repeat measurement on fixed markers in subsequent seasons. This type of geomorphological process study was greatly extended during the 1966 and 1967 field seasons. In the latter part of the 1965 field season, however, Martin concentrated on the moraine systems of inner Clyde Inlet and along the Clyde River valley to Generator Lake, dammed against the southern margin of the Barnes Ice Cap.

I was able to visit the field area between July 4 and August 31, at which time I made a reconnaissance of

Ekalugad Fiord with Olav that led to a decision to extend field research to that area the following summer. At the same time, Olav took me to inspect his critically important Cape Aston delta sites. I also made short visits to both John's and Cuchlaine's camps.

There were several reconnaissance flights that season with David in the helicopter, with the purpose of investigating whether or not the highest summits had ever been eroded, or at least submerged, by the continental ice sheet at its maximum extent during the ice ages. It has proved a long-running controversial topic in the realms of glacial geomorphology and vascular plant history, beginning with a dispute between biologists and geologists in Norway as long ago as the 1890s. One special difficulty still facing us in the 1960s, and for some time thereafter, was our inability to obtain absolute dates from high mountaintops that retained no material for C14 dating (the anticipated ages were far beyond the limits of lichenometry).

In wrestling with this problem, I realized that the area around Inugsuin Fiord provided helpful insights for estimating the age of apparent glacial erratics on mountaintops and of contributing to the controversy of survival of vascular plants throughout the ice ages. We were surrounded by summits capped by very thin patches of ice (small ice caps), which were presumably frozen to the underlying ground. As these patches melted, there would be no trace of their former presence. It was apparent that, given the similar yet more extensive situation of mountaintop glacierization during the ice ages, subsequent disappearance of thin summit ice cap cover would similarly leave no trace of its former presence. So, perhaps the main task should be to concentrate on searching for the upper limits of actively eroding ice. This would imply thick ice—so thick that it would not be frozen to its bed but rather would be in motion and capable of eroding the surface. At least this would provide an approximation of the maximum thickness of the major ice sheet during the last ice age, although there still remained the conundrum of the occasional

high-altitude erratic block within the zone that apparently had not been moulded by moving ice. At the same time, it appeared that a complete answer would have to wait for the development of techniques for dating the erratics and tors directly. This simplistic expectation proved too optimistic, as demonstrated by very recent developments (Margreth, Gosse, & Dyke, 2014). (A full explanation is reserved for chapter 11.)

Despite this obstacle, it was a project that I would take up in earnest in 1966. The 1965 summer, by contrast, enabled a first reconnaissance of the Baffin Island northeast coastal mountain situation. On another dimension, it laid the foundations for a close partnership with our helicopter pilot, David Harrison.

Building confidence in helicopter capability and safety

I had first met David Harrison in Ottawa when Olav and I were reviewing helicopter charter tenders earlier in the year. One of the most competitive bids for the contract had come from Spartan Air Services, an Ottawa firm with a long history of aerial survey and supply work on the original DEW Line. I took the precaution of meeting the pilot whom Spartan had proposed for the operation in the event that they won the contract. David's personality was immediately arresting, as were his experience and educational background. He had earned a Cambridge degree in economics before spending two years in the RAF, training under the RCAF as a NATO jet pilot. In civil aviation, he had flown helicopters on agricultural operations in the United Kingdom, Central America, and New Zealand; recently, he had specialized in bush and mountain flying in British Columbia. He was currently teaching geography at a college in Ottawa and flying helicopters for Spartan during summer vacations. Meeting David was like coming across an old colleague; sparks flew and there was a sense of instant rapport. I pressed my head office contacts to scrutinize Spartan's bid carefully, taking into consideration the background of the proposed pilot. Thankfully, this succeeded. David was to spend the next three summers with us on Baffin Island.

My most intensive operations with David and the helicopter were in 1966 and 1967. In 1965, I was still struggling to hold down my responsibilities as director of the Geographical Branch while also putting in a fair measure of fieldwork on Baffin. My close collaboration with Olav was essential in this, and during the summer of 1965 I was able to get away from Ottawa for most of July and August. This provided the occasion, from the impressive new base camp at the head of Inugsuin Fiord, for me to develop outstanding working relations with David.

The registration letters of the helicopter we used in 1965 were CF-FCK and this always—and not surprisingly—produced a degree of amusement. David proved a very special helicopter personality, and he was particularly interested in one of my objectives: to obtain wide-ranging photographic coverage of the exceptional glacial landforms, glaciers, and mountain and fiord landscapes of northeastern Baffin Island.[3] He also welcomed my interest in examining the summits of a large number of isolated peaks as a new challenge for his mountain flying techniques.

That spring through the Geographical Branch I had arranged for the purchase of a high-performance, medium-format camera (Hasselblad 500C) with several extra lenses. The arrangement of removable camera backs that permitted both colour and black-and-white photography through first-rate Zeiss lenses was one of its particularly attractive features.[4] And to facilitate the camera's most effective use, David was more than willing to fly with the passenger-side door removed so that I could avoid reflections from the helicopter's Plexiglas bubble and side windows. Furthermore, with his extensive experience and considerable knowledge of physical geography (one of his teaching subjects), he could quickly place the helicopter in an

ideal position for obtaining the best perspective for shutter release once a target had been identified.

Before any helicopter photo-run, however, I had to undergo David's "training for active passengers." First, he insisted that he had to convince me that his helicopter was safe—and through me, presumably, the entire team. To impress this upon me he claimed that FCK was safer than any "fixed-wing" aircraft; in fact, he maintained, with a twinkle in those steely blue eyes, that fixed-wing flying was downright dangerous in comparison.

David took on the attitude of a perfunctory instructer, explaining that I was required to undergo a series of "lessons" before I would be allowed to enter the rarified atmosphere of serious helicopter flying.

Lesson one. The Bell G2A was a small helicopter capable of taking either one passenger and a limited load strapped on racks attached to the top of the floats or two passengers in a pinch, but with only a modest load in the sides. In the event of an engine failure, it was possible to execute what was referred to as an autorotation. Essentially, the emergency response was to manually disconnect the engine from the main overhead rotor blades, rather like "de-clutching" the manual gears in a car, and doing the equivalent of freewheeling. As the main blades continued to "windmill" at the same rpm as normal, and the tail rotor was interconnected to the freewheel, the helicopter could still be steered and directed to a certain extent—though with two significant exceptions: an inescapably rapid descent and only one chance at the landing (no "going round again"). Enough momentum is built up in the descent to provide this one final "cushioning" of the eventual forced landing. Helicopter pilots in training, and occasionally out in the field, can practice this emergency manoeuvre by throttling back the engine and descending in autorotation, but recovering well clear of the ground. David explained much of this to me on the ground, and once we were used to flying together, he asked if I'd like to have a live demo (but with the engine still running, of course). I was excited at the prospect and agreed.

We were at a much higher level than the usual cruising altitude and directly above base camp, probably about a thousand metres. Grinning, David asked if I was ready; he then calmly throttled back the engine to idling and pushed down the collective control stick to disengage the rotors and force the freewheel autorotation. The extensive gravel terraces surrounding the Inugsuin base camp stretched out below us. There was also some open water in the mainly ice-covered fiord that could be used if necessary, as the machine had rubber floats and so was in that sense amphibian—provided it landed upright.

It was a great thrill. With my door detached and the air whistling past us, at a safe height the engine was brought into play and we circled in for a normal perfect landing.

Lesson two. To further increase my confidence as a passenger in cloudy weather, it was necessary to fly much higher than for the first "lesson": in fact, right through gaps in the layer of stratocumulus cloud that had its base about one hundred metres above the neighbouring mountaintops. First we soared up through one of those gaps to emerge into a wonderland of sun and blue with a startling white expanse below us. (Fig. 23) Then we found another gap in the cloud cover, disengaged the engine, and descended smoothly and rapidly through the cloud base to see the rocky pinnacles rushing up to meet us.[5] As before, there followed a gentle landing.

Lesson three. Several flights and routine landings later, David asked me, "Now for lesson three; what do you think?"

To which I replied, "Splendid. Are we going to loop-the-loop?"

David assured me this was not only mechanically impossible but also of absolutely no redeeming moral value. We boarded the chopper and coasted down the fiord to its midsection, where the mountains were

highest and their peaks sharpest. We landed gently on the top of one giant rock summit that rose about 1,200 metres above the deep blue-green water of the fiord with its subparallel lanes of broken ice. What a spectacular camera station! Out came the Hasselblad, and I took a round of both colour and black-and-white photographs while David lay out full length in the sun on the bouldery rubble of the mountaintop.

"What now?" I asked.

David simply answered, "Wait and see!"

So we got back in, buckled up, and he started the engine. After some minutes I realized that we were not able to take off. He assured me that this had not been planned, so we got out and looked over the edge. David commented, "Looks like it would be a difficult walk back to camp, but if we did climb down, how would we rescue my helicopter?" He casually off-loaded the two five-gallon drums of extra avgas that had been strapped onto the floats, and in a carefree tone said, "I can retrieve them when I am on my way back next time without a passenger." (At this point a dark thought entered my mind.) David continued, "As it is now, at this altitude and with you on board and the gas on the side, our engine just doesn't have sufficient power to lift off."

When we restarted the engine, David pulled the helicopter into the hover "cushion" about ten metres above the summit and edged sideways until we were over the fiord, with a sheer wall of granite falling about a thousand metres straight down on one side of us and open space on the other, David casually disengaged the rotors. We seemed to be making the most rapid mountain descent of my life, and the granite wall appeared to be at arm's length. It was a very stimulating lesson and I was now fully confident of the ability of David to take me safely into and out of potential research locations, places that often could not be reached any other way. Back at camp, in calmer retrospect, I understood how David—realizing that he would be required to make numerous mountaintop landings that might well be scary for

Fig. 23: "Above the clouds." As the chopper rose through the thick layer of cloud we were greeted by high mackerel cirrus and a blue and white wonderland. (Photo: July 1965)

a passenger—was aiming to ensure that I would be relaxed and not likely to panic or jump out.

So began for me the first of three summer field seasons combining successful photography and geomorphology research with a frequent sense of adventure. Only once did my pilot seem to have any misgivings about our mountaintop perching and hovering for perfect shots ("Jack! You've turned me and my helicopter into just another accessory to your damned Hasselblad!").

Expanding research all the way from Foxe Basin to Baffin Bay

The 1965 summer on Baffin Island also witnessed a major expansion in glaciological and glacio-hydrological research to match Gunnar's activities in the Canadian western mountains. Complementing Brian's extensive work on the Barnes Ice Cap, Gunnar spent a couple of weeks between the Lewis Glacier and River and the Decade Glacier above the Inugsuin Fiord base camp. This involved additional training for Mike Church and Bill Rannie, who along with several other students were to take full control of the attempts to determine the flow and silt content of the highly turbulent Lewis River. This group included yet another first-year undergraduate, Barry Goodison. Also in May, Gunnar, assisted by Chris Bridge, set up a camp high on the Decade Glacier and began the detailed task of mass balance, climate, and hydrological studies. Chris was left in charge while Gunnar transferred operations from Baffin, hurried to Ottawa, and then headed out west to his newly selected glaciers in Alberta and British Columbia. Barry Goodison's early Baffin experience involved assisting Mike Church in his glacio-hydrographic study of the Lewis River.

From the end of May, Chris Bridge and Bill Rannie took care of the Decade Glacier mass balance and meltwater runoff measurements. A meteorological station was set up and observations maintained from June 14 until August 20. Despite the limited observations, it was apparent that the 1964–1965 year had proved a slightly positive one for the IHD-selected glacier. Working conditions were vastly improved when a helicopter lift added a snowmobile and a "Gunnar-type" A-frame hut, the latter to be firmly anchored on the northeast margin of the glacier at 950 metres asl. Several of the undergraduates were flown in to provide assistance, in accordance with the general plan to transfer students to difference phases of the overall operation so that they received as broad a training experience as logistics allowed.

For the first time, Baffin Island operations extended from the islands in Foxe Basin across the entire width of the main island, the Barnes Ice Cap, and the mountains and fiords of the northeast to the outer coast fronting Baffin Bay. This level of activity was possible because we had been able to justify the large increase in annual budget necessary to ensure a worthy Canadian glaciological contribution to the IHD. In this sense, the 1965 season could be regarded as the pinnacle of my aspirations both for Baffin Island research and for the Geographical Branch itself. Little did I realize at the time that I was due for a major letdown and that the IHD would prove the beginning of disintegration for the branch as a result of a forced departmental reorganization.

All personnel were back in Ottawa by the beginning of September. Yet another season had been marked by outstanding collaboration and enthusiasm among all expedition members, as well as between them and the aircrew. DEW Line assistance had remained at a high level. The logistical importance of Fox-2 and Fox-3 stations could not be exaggerated. The early record of neither illness nor injury had been maintained. Cuchlaine King and her two assistants, Jane and Wendy, had had a very enjoyable and productive summer and were impressed by the courteous and enthusiastic reception they had received from the DEW Line personnel. I felt much satisfaction in being able to report this to the senior departmental members who had put up so much opposition to the proposal that women deserved the same opportunity as men to conduct research in the Arctic.

Ottawa developments and planning for the 1966 field season

This critical winter of 1965–1966 not only found me committed to preparations for what was to become, yet again, the largest and most complicated Baffin Island field season; I was also heavily involved with general Geographical Branch administrative affairs. These included continued efforts to expand the economic geography activities. Dr. Michael Szabo was recruited as chief of the Division of Economic Geography, and a significant study of Prairie railway branch line optimization was initiated under the leadership of Tony Burges. The desk atlas and the reinvigorated National Atlas program moved ahead under the direction of Gerry Fremlin. And the not-so-routine affairs, such as meetings of the newly created National Advisory Committee on Geographical Research and of the Canadian Permanent Committee on Geographical Names, required special attention. There were also discussions on Canada's hosting of the 1972 IGU Congress, much of the expense of which would have to be covered by the branch.

Much work was put into compilation of a special glaciology issue of the *Geographical Bulletin*. This involved analysis of the large amount of field data, from both Baffin Island and the mountains of Alberta and British Columbia. Thus George Falconer, Olav Løken, Gunnar Østrem, and Brian Sagar were all heavily involved, with a notable assist from Ross Mackay of UBC. The issue, which included Canada's first published regional map of glaciers (southern Alberta and British Columbia), was what I considered a critical undertaking in order to demonstrate that the branch's glaciological program was rapidly producing substantial results. It was published by the Queen's Printer in early 1966 as *Geographical Bulletin*, volume 8, number 1. I regarded the production of

the special glaciology issue as a showcase of what the branch could do in glacier research, demonstrating the progress from rapid and effective performance in the field all the way to a peer-reviewed journal publication within a rather short time.

The most time-consuming but politically important development was the recruitment of members for the newly approved Canadian National Advisory Committee on Geographical Research. The committee's inaugural meeting was held on November 5, 1965. I was thrilled with the highly successful results.[6] While most members were chairs of geography departments from across Canada, I was also able to attract senior members of industry (for instance, from Imperial Oil, The Tower Company [Arctic construction], and British Newfoundland Corporation) and other federal and provincial representatives. Dr. van Steenburgh had originally agreed to chair the meeting although, being indisposed, Dr. James M. Harrison, recently promoted to the position of assistant deputy minister, delivered the address of welcome on his behalf. Dr. Harrison forecast an important role for the advisory committee in shaping departmental and national policy regarding geographical research. He outlined changes taking place in the Department of Mines and Technical Surveys, attendant upon creation of the new Waters Research Branch. I gave a report on recent research and organizational developments in the Geographical Branch and welcomed the creation of the advisory committee as an important instrument for both strengthening collaboration with academic geography and finding ways for more effective relationships with Canadian industry.

It was rather conspicuous that all attending the meeting were male, with the exception of my recently recruited and extremely able chief administrative officer, Alexandra Cowie.[7] This inaugural meeting proved extremely convivial and promised to be of great importance for the continued development of the Geographical Branch. The concluding recommendation made to the minister was to strengthen both economic and physical geography within the branch rather than risk negative effects due to reorganization of the department. A grants-in-aid subcommittee was set up and, for the 1966–1967 period, twenty-three modest research grants were awarded to academic geographers throughout Canada. A second meeting was scheduled for the following February.

The most serious development of the winter, however, had to do with the start of the IHD. Responsibility for water research, both freshwater and marine, and its management as a vital Canadian natural resource had traditionally been divided among several federal government departments. Competition for "control"—and hence very much enlarged annual budgets—between the Department of Northern Affairs and Natural Resources and our Department of Mines and Technical Surveys, was becoming especially pronounced. Yet while glaciers are conspicuously part of the hydrological cycle, I had not initially suspected that this would draw the Geographical Branch into departmental and interdepartmental politics; my vision had instead been that it would provide a base for fruitful collaboration. The next twelve months were to prove me wrong.

In the briefest terms, our department was reformed as the Department of Energy, Mines, and Resources, while all water-related affairs were placed under the responsibility of a newly created Inland Waters Branch, to which my greatly cherished Glaciology Section was transferred. This had no effect on the 1966 Baffin field season, nor on the Glaciology Section in the short term. Everything remained in place—but very shortly it became apparent that a significant number of branch staff would be reporting to Dr. Al Prince, the newly appointed director of the Inland Waters Branch. I discussed my concerns with Dr. Van, who explained that arrangements were moving beyond his control, especially with the approach of his sixty-fifth birthday and, with it, compulsory retirement. A new deputy minister (DM) was to be appointed, and that might (or might not) have repercussions all the way down to the Geographical Branch. However, Dr. Van urged me to continue with what I was doing. In retrospect, this augured the end of the Geographical Branch, not that I realized it at the time.

These uncertainties, as unnerving for me as can be imagined, could be submerged beneath a sense of achievement for the field program and the personal satisfaction of realizing that a small helicopter, with the right pilot, could achieve a level of effective research combined with high adventure that I could never have imagined before experiencing that first autorotation. It was a feeling that has stayed with me for a lifetime. Meanwhile, I knew that a great deal more was in store for 1966 and 1967.

Fig. 24: Inugsuin Fiord from above base camp. The fiord is still impenetrable to our small boats. The sandy margin fronting the fiord in the foreground shows a series of raised marine terraces carefully excavated by Olav Løken in 1964, which enabled him to date the progressive emergence of the land by radiocarbon dating a collection of seashells. (Photo: July 1965)

❧ SUMMIT EXPERIENCES ❧
AND EAST COAST RESEARCH, 1966

Our helicopter pilot David Harrison made a remarkable contribution to the final three summers of the Baffin Island expeditions (1965, 1966, and 1967). He imperceptibly became a full member of the team. His previous flying experience in a variety of fields, from agriculture and forestry to oil exploration, in many parts of the world, together with his extensive landscape curiosity and interest in our research, gave him a special position as confidant, far beyond his official duties as helicopter pilot. He was perceived by students and regular staff alike as an independent-minded, wise, and sympathetic friend who was outside the formal federal government order of command. Tom, his engineer, was a stolid, reliable, mechanical genius, who reinforced everyone's vital confidence in David's ability to get people safely into and out of difficult terrain in almost all types of weather. Tom was also an excellent cook, and his fresh hot bread became a memorable addition to the somewhat routine freeze-dried menu.

David was enthusiastic about his mission and always ready with exciting stories about flying in the Canadian West, New Zealand, and beyond—in both accessible areas (flying Santa Claus at Christmas in New Zealand; doing airborne traffic patrol over Ottawa at rush hour) and far-from-base remote situations (delivering a single-engine monoplane from Croydon, UK, to Rangoon for the Burmese Air Force; overflying dense tropical rainforest in Panama). He had published several articles, which were embellished with his sharp and still-British sense of humour. Three separate articles in *Shell Aviation News* (Harrison, 1967a, 1967b; Harrison & Benjamin, 1967) with titles that speak for themselves—"Relief Information Unreliable" and "Peak Performance"—were based, in part, on our Baffin Island expeditions and I was present on several of the adventures he described. They did not refer explicitly to our operations; I suppose we stretched too many air regulations. David also co-authored with me a colourful article entitled "Rotorcraft on Research," in the *Canadian Geographical Journal* (Ives & Harrison, 1967), which spoke to the more serious logistical issues of Arctic terrain research.

David's duties in Baffin were twofold. First, he had the task of flying in logistical support for a large and complex field research party in a remote mountainous setting amid the usual vagaries of "summer" Arctic weather (snow, cloud, fog, rain, high winds). The title of the first, two-part article, "Relief Information Unreliable," relates to some of the challenges faced by pilots using aviation maps that had not yet integrated the relief information from the quite recent air photos (see chapter 3, note 9). Instead, the aviation maps had just those three stark words of warning—"Relief Information Unreliable"—in place of the usual contours and peak elevations. To this should be added *communication systems extremely unreliable*, another endemic hazard of the period for bush aircraft in the Far North. For us, the safety factor was enhanced by addition of a second helicopter in 1967, chartered jointly with the Canadian Wildlife Service, members of which were researching the summer nesting grounds of the blue goose in the Nettilling and Amadjuak lakes area, in southwestern Baffin. Similarly, late in each summer season, as more open water became available, the arrival of a Cessna floatplane added to the flexibility of our logistics and enabled gas-caches for the helicopters to be laid out relatively inexpensively. The Cessna was also a good emergency backup, although fortunately it was never needed.

The helicopter's first task for the 1966 summer was the routine ferrying of arriving and departing personnel between the nearest DEW Line site, the Inugsuin Fiord base camp, and scattered field sites, plus the frequent transfer of two- or three-person light tent camps across an area of about 250 by 350 kilometres. Its second task was direct support of specific research activities. This included placing expedition members at numerous locations along the outer northeast Baffin Bay coast—north and south of Clyde River, where open water for the Cessna floatplane was either non-existent or uncertain. An important part of the fieldwork entailed the systematic tracing of glacial and raised marine shore features over long distances and support of the high-altitude glaciological work on the Decade Glacier. To this must be added my field projects, such as the examination of numerous mountaintops for evidence of past glacial (ice age) activity and photography of the entire fiord and mountain landscape using the Hasselblad medium-format camera. David assured me that this combined project was his favourite because it gave him a novel and unusual opportunity to contribute to extending the frontiers of Canadian Arctic mountain geography.

The larger logistic problems facing our field plans related primarily to remoteness from Ottawa, or "the South," and our dependency on access via the DEW Line. The Inugsuin base camp itself was located 120 kilometres from the nearest DEW Line station (Fox-3). The challenge to operations in northeastern Baffin Island was directly associated with the landscape—a 160-kilometre-wide mountain zone cut by numerous precipitous fiords, themselves exceeding 100 kilometres in length. Between the fiords, the long rugged peninsulas support numerous local ice caps and glaciers, some extending down to sea level, and are dissected by deep, glacially carved valley troughs. The outer coastal strip is mostly flat and, in some places, ten to fifteen kilometres wide; in others, it can be a narrow strip of a hundred paces. The coastline is backed by a steep rise to alpine-like peaks, many of which extend to heights of 1,200 to 1,500 metres asl. The highest summits, forming the height of land, cut at right angles across the midsection of the fiords. From that point, they slope gradually down to the southwest, with the relief becoming less rugged, until merging with the rolling inland plateau surmounted by the Barnes Ice Cap.

The only settlement in the entire field area was Clyde River, on the site of a traditional Inuit encampment; by the 1950s, it had an RCMP post, radio station, and small Hudson's Bay Company store. Some nine kilometres farther northeast, on the outer coast at Cape Christian, there was a U.S. Coast Guard LORAN station not part of the DEW Line system. The topographic map coverage, at scales of 1:500 000 and 1:250 000, was totally inadequate for secure navigation from the air. We were aware of the height of some of the highest summits as reported by the mountaineering exploits of the AINA 1950 Baffin Island expedition (such as Eglinton Tower at over 1,300 metres). However, detailed topographic information was scant and, aside from the major fiords and the more prominent headlands, even place names were lacking. As a helicopter pilot in a rugged mountainous region that was also subject to sudden changes in weather, especially rapidly spreading fog in the outer fiords and along the coastal lowlands, David had to make sure that he could find his way "home" (usually with one or more of us on board) if suddenly hit by inclement conditions—and without running out of fuel.

Sudden changes in wind speed and direction provided a special challenge. Winds blowing in opposite directions could be encountered at certain altitudes, especially near the steep cliffs of the fiords, where the helicopter could suddenly and alarmingly gain or lose more than a hundred metres of altitude in the updrafts and downdrafts. Moderate to strong katabatic winds associated with the ice caps and glaciers required a "seat-of-the-pants" approach to flying. The only available weather forecasts were those we could cook up ourselves. These improved as we grew progressively more accustomed to the region, but local conditions were something David usually had to anticipate and deal with based on his day-to-day flying experience.[1]

Of great advantage, being close to latitude 70° N as we were, were the twenty-four hours of daylight extending into August that meant there was no risk of being "benighted." Midnight operations were common as it was necessary to make use of all available good weather, and working by the light of the midnight sun was delightful. Some of the most unforgettable field research moments of my life involved writing up notes while sitting with David on a mountaintop close to midnight, the helicopter quietly parked and out of sight behind us, an abrupt drop in front of us of more than a thousand metres to a dark green fiord, with fiery red pinnacles and ice caps stretching across the far horizon, and the all-pervading silence.

Exploration by helicopter along Inugsuin Fiord

One morning's account of that summer may provide the reader with a sense of the outstanding advantage provided by field helicopter operations. In perfect weather, it was to be a rewarding day in all aspects. We were all set to leave base camp by 9:00 a.m. David

and his engineer Tom had checked and re-checked everything that could be checked. The passenger-side Plexiglas door had been removed and I sat next to David, firmly belted in, with the large Hasselblad case gripped between my knees and the camera with its 250mm lens clasped in both hands. We had lunch, extra food, a light tent and primus stove, climbing rope, and ice axe all carefully strapped onto the external cargo racks.[2] (Map 8)

Soon we were climbing up past the Decade Glacier (Fig. 25) and heading north-northeast in calm weather with strands of cirrus, high above us, reflected in the still fiord waters. Twenty-five kilometres into our flight, we turned the first great bend of the fiord and passed under the Inugsuin Pinnacles (officially named Nuksuklorolu Mountain on the 1984 McBeth Fiord 1:250 000 map) at about six hundred metres asl. As we approached a second 90-degree bend, we headed for a spectacular trough stretching to the southeast. I indicated to David that we should leave the fiord and fly through the trough. We flew past glaciers large and small that were descending from ice-capped summits all around us. Another 90-degree turn led us into what was later named Perfection Pass. (Fig. 26) There were several lakes, one of which was about three kilometres long and partially dammed at its far end by a series of end moraines. The innermost moraine was steep-sided and very prominent. It formed the margin of a glacier that jutted onto the floor of the pass. The outer moraine was much more subdued in shape and partially covered with tundra vegetation, indicating that it had been laid down probably several thousand years ago.

The pass ran roughly parallel to the main fiord but was separated from it by a massive rock wall capped by ice. The pass opened onto an area of gently rounded hills, with views across the outer part of Inugsuin Fiord where it begins to merge into Clyde Inlet. At this point, we turned sharply westward and flew back along the fiord until we were close to our original entrance to Perfection Pass. We then gradually ascended the face of a high mountain that I later

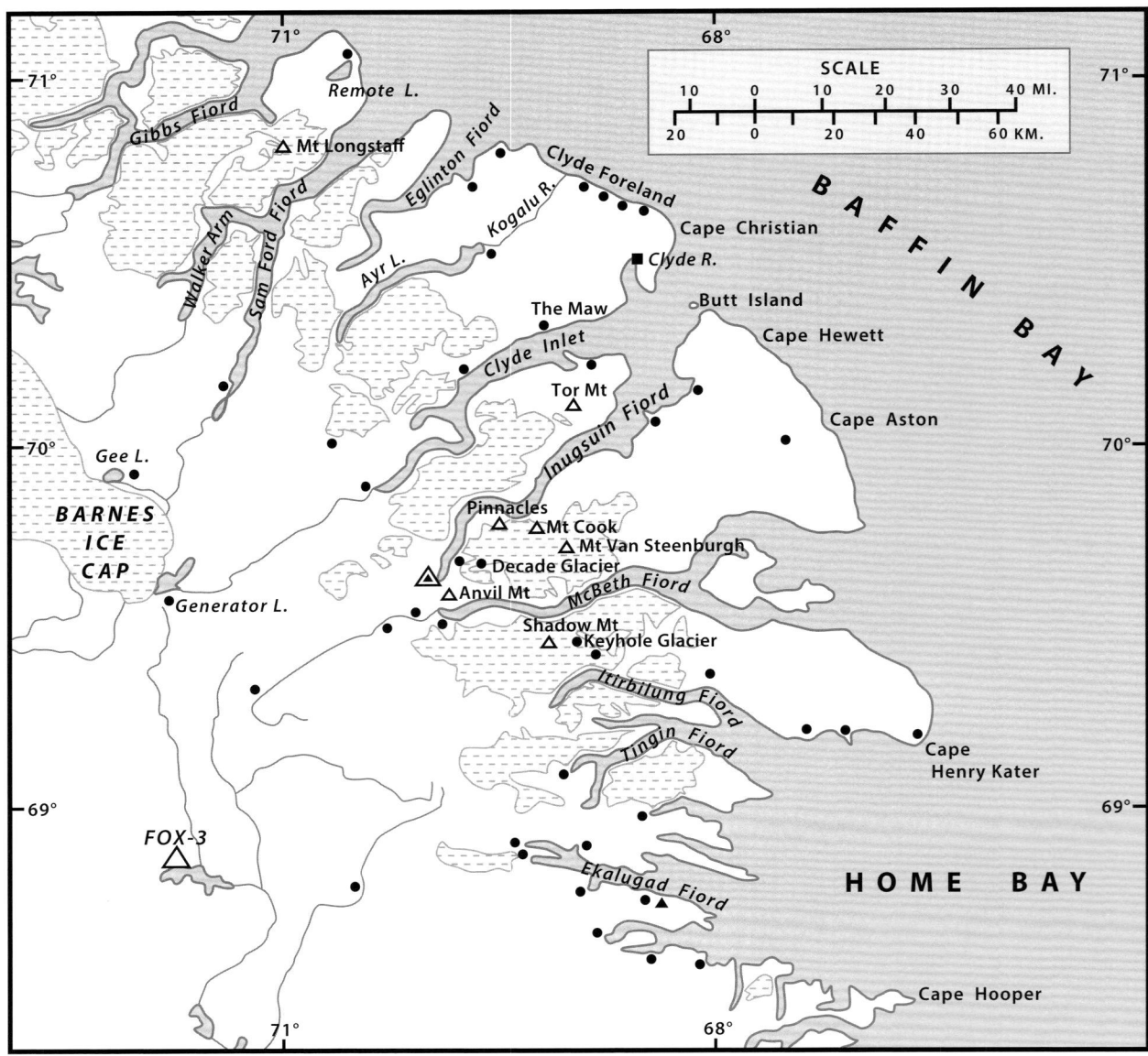

MAP 8: Detailed coverage of main activities in the fiord and outer coast in 1966 and 1967. Several of the names proposed for previously unnamed mountains and glaciers are included. The double triangle represents the Inugsuin Fiord base camp.

named Mount Cook in memory of Frank Cook, an old friend and member of the Geographical Branch staff who had died in tragic circumstances two years earlier. We landed on its broad summit at about 1,200 metres and switched off the engine. It was time for lunch and an extensive photography session.

The broad summit was surrounded on all sides by precipitous rock walls and partly covered by a

small ice cap. By far the best photo angle was to be obtained by leaning out as far as possible over the northeastern edge. I found that I could not lean out far enough and maintain my balance so I anchored myself with the climbing rope to David and the helicopter, which served as a useful belay with handy places for attaching the rope. The security of the belay enabled me to edge out and down a smooth sloping

Fig. 25: Decade Glacier, named for the International Hydrological Decade. This late-evening view of the glacier represents a perfect archival record. The rose-tinted light-toned areas, a result of limited rock lichen growth, had been covered by permanent ice and snow more than one hundred years previously. The progress of the ablation season on the glacier displays the exposure of different years of ice as well as the small remnant of snow at the higher altitudes from the previous accumulation season. Old end moraines in the foreground reach into the shadow. (Photo: August 1966)

Fig. 26: Perfection Pass. The terminus of the glacier, which has penetrated to the floor of the pass, has retreated from two sets of end moraines of greatly different ages. The multiple ridges of the inner moraine system are light in colour because of a near total absence of vegetation. They are ice-cored and unstable. The outer low arcuate form (concentric to ithe inner moraine must be several thousand years old; its former ice core has melted and vegetation has spread across its surface. (Photo: July 1966)

Fig. 27: View from Mount Cook. The view shows the precipitous south face looking toward the outer section of Inugsuin Fiord. Baffin Bay is in the distant haze. (Photo: July 1966)

Fig. 28: A textbook example of glacier terminal features, from the summit of Mount Cook: lateral, medial, and end moraines, eskers, moraine-dammed lakes, and glacial outwash plain (sandur). (Photo: July 1966)

Fig. 29: Helicopter descent for closer inspection of glacier features seen from the summit of Mount Cook (see Fig. 28). (Photo: July 1966)

rock surface to reach the desired camera angle. David was conspicuously uncomfortable! When I was back on level ground, unfastening the rope from the helicopter, he insisted I had forced that manoeuvre on him as payback for the adventure of demonstration autorotations and other "passenger training" exercises in the helicopter the previous summer. However, the outstanding photograph justified—for me, at any rate—my true and very proper motivation.[3] (Figs. 27, 28, 29)

On another occasion the following week while I was fully engaged with the Hasselblad on a summit similar to that of Mount Cook, a somewhat bored David inadvertently nodded off to sleep beyond my line of sight and rather too close to the edge of a sheer precipice. Meanwhile, I completed the photography and replaced the various pieces of the camera system in its case. I looked around, but my pilot (and my only way off the peak) was absolutely nowhere in sight. My heart leapt and I began an agitated search. I quickly found him asleep behind some large boulders but, for a sleeping body, I thought he was far too close to the edge of disaster. I approached very gingerly, found a secure foothold, and took a firm grip on his collar, gently nudging him awake. His comment was this: "You might have had to make an embarrassing call back to base camp—that is, if the radio ever worked." (In retrospect, I'm sure David was really feigning sleep, but at the time it was all too believable.)

After lunch, we made landings on several mountaintops on both sides of this section of the fiord, always searching for glacial erratics and any other indications of the former passage of the Laurentide Ice Sheet. Finding none, we landed on the lower slopes of one of the two glaciers down onto which we had looked from the top of Mount Cook the week before. (Fig. 30) The objective was to examine and photograph the details of its terminus close to the fiord margin. Following this, we decided that our final sortie of the day would be a landing on top of one of the Inugsuin Pinnacles. The first landing approach proved too

hazardous because of the updrafts and downdrafts and insufficient flat space on the summit, but David insisted on compensating for my disappointment by changing altitude and direction to complete a figure-8 flight around and between the upper parts of the two highest rock towers, giving me another remarkable photo op. As the gusty downdrafts seemed to have subsided, we made a new approach, to a possible landing spot on the edge of a prominent broad summit several kilometres farther in from the fiord. This massif, rising at least 1,300 metres, was almost entirely covered by a summit ice cap, leaving just a narrow ledge of bare rock and boulders along its southwest edge. (Fig. 31) We intended to land on the ledge and inspect the surface. But, as we descended to within about fifteen metres of the ledge, we felt ourselves again caught in the downslope air and were forced over the edge, dropping sixty to eighty metres before David regained complete control. As we pulled away from the influence of the downdraft, he casually asked, "Well, Jack, how about another try?"

Even with a different approach, we had the same experience of being in the grip of a mountain god, only more forcibly. Once more David pulled safely away. He asked again, "One more try?" (I suppose trying to gauge how important this particular place and its samples were to my research). To my positive response, he replied, "Oh dear, I was sure you would say no! Well, sorry, Jack, it's time for tea." And so we wisely called it a day and pointed back home along the fiord to base camp.[4]

Investigating summits from Anvil Mountain to Clyde Inlet

The helicopter had been employed for several days to make a series of camp moves while, back at base, I had prepared for a more extensive investigation of mountaintops from the head of Inugsuin Fiord to the coast north of the Clyde River settlement. With our usual pack of emergency equipment, we flew down

Fig. 30: David Harrison provides a scale. The eskers seen from above are ice-cored—this was the case for nearly all the glacier terminal features. In southern Canada such features emerged from the ice sheets of the last ice age and were not usually ice-cored because of the much milder climate; therefore, their complete form was largely preserved. In the case of Baffin Island and other regions under an Arctic climate, such ice-cored features would largely disappear with glacier retreat and a warming climate. (Photo: July 1966)

Fig. 31: "The Inugsuin Pinnacles," Nuksuklorolu Mountain. The pinnacles are illuminated by the late-evening light. After failing to make a touchdown with the helicopter, David attempted the ledge at the edge of the summit ice cap in the distance (shown by arrow). This attempt was also aborted, due to persistent downdrafts. (Photo: July 1966)

the fiord and ascended over its southern end, aiming for what would later be named Anvil Mountain. Its massive bulk rose almost to one thousand metres about six kilometres south-southeast of Inugsuin base camp. As we approached touchdown, we saw it was partially ice-capped although the highest point was a wide platform of frost-shattered bedrock, essentially featureless except for one giant boulder—the "anvil."

We landed close to the highest point, the huge boulder, which was about nine metres long and three metres high. It was composed of coarse granite gneiss. How had it formed? Was it a glacial erratic at a critical elevation? Or was it a more resistant section of the underlying bedrock that had projected higher and higher as its less resistant surroundings had been worn down during eons of surface erosion? It was situated in an area of large frost-shattered debris—the common European technical term is *felsenmeer* (sea of boulders). Identification of such large isolated blocks as glacial erratics had been a major source of controversy in Scandinavia, Greenland, and Labrador for more than sixty years.

Closer examination of the huge anvil-shaped block revealed a much smaller block sitting on its crest. (Fig. 32) It was composed of a contrasting type of granite gneiss. This small block must have been placed there by overriding glacier ice, even if the larger block on which it rested was *in situ* (i.e., in place as a section of the bedrock especially resistant to erosion—in other words, a tor). However, I thought this

Fig. 32: Anvil Mountain. The boulder field of the plateau-like top of Anvil Mountain is capped by a giant block (David Harrison is pictured for scale). While it was difficult to determine the origin of such a block—a glacial erratic or part of the bedrock more resistant to erosion than its surrounding rock?—in this case the answer was suggested by the small boulder precariously perched on its summit (shown by arrow). Moving glacier ice must have emplaced at least the small boulder. (Photo: August 1966)

episode of glacial history must have occurred long prior to the last ice age. This was a tentative conclusion, because in the 1960s there was no means of resolving such a problem by any known dating technique. The interpretation—"How old?"—was part of the related controversy. The discovery at Anvil Mountain was at least another step forward in unravelling the glacial history of the region, even if the precise chronology could not be resolved.

After shooting both monochrome and colour rolls of film, with David as a convenient scale, we set off northeastward along Inugsuin Fiord. We flew close to the northwest wall of the fiord, which, for most of its midsection, rises precipitously more than eight hundred metres from sea level. Where the slopes were less steep, short sections of glacial lateral moraine could be traced, matched by similar features on the opposite side of the fiord at about the same altitude. Both sets were about six hundred metres above the fiord and sloped gradually down toward the outer coast. Here it appeared certain that we were following the outlines of a former giant outlet glacier that, during the last ice age, must have drained from the continental ice sheet over central Baffin Island and flowed down northeastward to enter Baffin Bay. This episode would have been much more recent than the occasion on which the small boulder was set down on the summit of Anvil Mountain.[5]

We made several high-altitude landings as we progressed northeastward. These only produced exposures of more *felsenmeer* (mountaintop detritus), and there were no indications of glacial activity except for the many small present-day ice caps that covered much of the higher ground. After we had passed the Inugsuin Pinnacles, the summits became progressively lower and less alpine in character. At the outer part of the fiord, just before it merged with Clyde Inlet, the highest points were gently rounded and heavily mantled with the ubiquitous boulder debris. We touched down on one summit that was about 750 metres high with a very broken surface from which emerged a series of prominent tors (bedrock knobs) rising between three and five metres above the general level. According to the more conservative of the Scandinavian hypotheses, throughout the ice ages moving ice could not have reached as high as the projecting tors, otherwise it would have eroded them. This was a problem that I had encountered in the Torngat Mountains, where I had discovered definite glacial erratics sitting on top of tors, thereby confounding the original strict interpretation of Eilif Dahl (1955, 1961), a leading Norwegian arctic-alpine botanist and protagonist of the "nunatak hypothesis." Well below us, probably below the three-hundred-metre level, we could distinguish the continuation of the lateral moraines that we had been following since shortly after leaving the head of Inugsuin Fiord. Below them again, we could follow the glistening, ice-moulded, and polished bedrock surfaces—that is, unequivocal evidence of glacial erosion during the closing phases of the last glaciation. (Figs. 33, 34, 35, 36)

After photographing the tors and collecting rock and soil samples, I decided to name the location "Tor Mountain." We then headed across Clyde Inlet intending to land above the cliffs that rose abruptly to form the far side of the fiord. As we approached, we could estimate that the cliffs rose between five and six hundred metres to a gently rolling summit surface and our intended landing spot. The air was calm and the surface of the sea barely rippled. Stray pieces of pack ice were scattered below us, thickening in coverage toward Baffin Bay, which was solid white with loose pack interspersed with icebergs, all drifting slowly southeast along the coast. Conditions, at least from our vantage point, seemed ideal.

I remember a distinct sensation of excitement as we closed in on the Clyde cliffs. We were about thirty metres above them. Suddenly I felt we were in free fall. My first reaction was to catch the quite heavy Hasselblad case as it flew up toward the roof of the Plexiglas bubble.[6] In no time, far down the cliff, we shuddered to a complete halt like a sudden stop in an elevator. I thought we must have hit some rocky island. Then we began to climb, but also abnormally

Fig. 33: Mountaintop detritus. Extensive and deep expanses of frost-shattered bedrock had been used extensively in Scandinavia, Greenland, and Labrador as an indication that the high summits had never been submerged beneath a continental ice sheet. Thus, they could have provided locations for the survival of arctic-alpine plant species (the nunatak hypothesis). (Photo: August 1966)

Fig. 34: Mountaintop detritus on three distant summits and in the foreground. Note the small "bump" on the summit at right, shown closer in Fig. 35. (Photo: August 1966)

Fig. 35: The term "tor" originated in Dartmoor, in southwest England. Tors are weathering residuals of harder rock left upstanding as surrounding, less resistant bedrock is worn down more quickly (over a long period of geological time). Because many tors are small and could easily be obliterated by the motion of powerful glaciers, they have often been associated—together with mountaintop detritus and weathering pits (see Fig. 85)—as evidence of the former existence of nunataks. However, this interpretation is challenged by the discovery of glacial erratics on top of tors. (Photo: August 1966)

Fig. 36: Closeup of a tor with scale.

rapidly. To my dismay, I realized that David was transmitting the international distress signal—"Mayday, Mayday, Mayday"—followed by "Cape Christian, do you read me?" (The Cape Christian U.S. Coast Guard LORAN base was the nearest place with twenty-four-hour receivers likely to hear us.)

According to David's subsequent assessment, our first precipitous descent had been "at a rate of 800 feet/minute which seemed to be leading us straight to a watery grave and then, after two sledge-hammer jolts, into an equally violent 1,500 feet/minute updraft." He went on to recall what happened next:

> At times like this, though one is almost too busy to be terrified, it is nice to have radio contact with someone. I passed my position and plight to . . . Cape Christian. Back came a reassurance in a somewhat incongruous Alabama drawl: "Cap'n says to come raht in, sah. He'll have a cool beer waitin' foh ya if ya can make it in, sah."

With mixed emotions, we failed to reach the beer, which would have felt so good in my dry mouth even though I rarely drink beer. After one slightly less threatening repeat of the "big dipper" performance, David struggled out of the turbulence and fled to the south side of the fiord. Here we set up our emergency Arctic Guinea tent in absolutely calm conditions. Wind? What wind? The low Arctic sun was beaming down from a near-cloudless sky. A life-threatening situation had been a highly localized event. (Fig. 37)

This time, the radios worked to great effect. David was quickly in radio contact with Inugsuin base camp, from which engineer Tom's step-by-step instructions walked David through a thorough physical "external" of the entire craft. To David's surprise, a piece of the metal framework had actually snapped off, but apparently that was of no serious account, merely a measure of the immense forces of nature that had held us momentarily in their grasp. David's and Tom's main concern was the alignment of the main blades and the tail rotor. Together they determined that there was only a slight distortion on the main blades, which Tom thought would not cause any problem. In our otherwise perfect Arctic paradise on the remote shore we rested a calm twelve hours at our campsite and cautiously flew back to base camp the following day. The section of the Clyde Inlet cliffs where we experienced our encounter with the ferocious downdrafts was subsequently named The Maw, and the small island at its base Downdraft Island.

During our long wait on outer Clyde Inlet, David related another cautionary tale. The previous week, he had helped set up a light camp on Ekalugad Fiord from which John Andrews was to make a more detailed examination of the local geomorphology than Olav and I had managed the previous year. Leaving John and his field assistant, John England, David had taken the helicopter out over Baffin Bay beyond the DEW Line station on the summit of Cape Hooper, at about six hundred metres. He had found himself instantly enveloped in a thick fog. Rising through it, he could make out the mountain summits beyond the fog-enshrouded coast. As he was not enthusiastic about spending a lonely and very cool night on a mountaintop, he eventually found a hole in the fog and descended, thinking that he could get underneath it. In this he was mistaken. The fog was socked in tightly down to almost zero-zero altitude/visibility on the landfast sea ice. So he cut forward speed down to little more than a crawl, feeling his way, with some concern, toward the coastal cliffs. As he crept along, close to head height above the sea ice surface, all of a sudden something and someone loomed up immediately ahead of the bubble. David practically crashed head-on into two Inuit on snowmobiles. He related that it would have been hard to determine who was the more astonished; he made a quick decision not to stop, as he now realized that he could follow the snowmobile tracks around Cape Hooper and back into the fiord. A few more anxious minutes later, he emerged from the fog and found his way back to camp. He told John he would have loved to hear the Inuits' version of the event.

Fig. 37: SOS to Inugsuin Base Camp. The helicopter sits awaiting a radio checkout from engineer Tom at base camp. David had managed to squeeze out of the massive downdrafts under the cliffs on the skyline. Fourteen hours of peace and relaxation followed before a gentle cruise back "home." (Photo: August 1966)

Shadow Mountain and "The Keyhole"

Several helicopter traverses were made along McBeth and Itirbilung fiords, the next two fiords south of Inugsuin, and the same pattern of mountaintop hopping was continued. From the great bend of Itirbilung Fiord, about fifty kilometres out from base camp, we came upon a massive U-shaped trough trending roughly north–south. It cut through a series of the higher mountains that were dissected by several large glaciers descending onto the trough floor. One particular mountain seemed especially appropriate for my purpose. It was composed of three separate summits, the main one looking as if it was one of the highest in the entire area. The top was bare of ice and snow, and we made an easy landing in good weather. It proved an excellent survey point. And there was no evidence on the summit of the former passage of glacier ice.

From our summit perch, we could see far below us a large glacier with two gigantic erratic blocks on its surface. This afforded me an outstanding, textbook example of glacial erratics actually in the process of down-glacier transport, eventually to be deposited on the trough floor once the glacier had carried them to its terminus. We flew down. David was in an expansive mood and landed on the very top of one of the blocks. He asked if I would go out onto the glacier surface and photograph the helicopter sitting there on its precarious-looking eyrie. There was a small snag. I had forgotten my climbing rope and, since the sides of the block were sheer and more than six metres high, we were marooned on top. So we lifted off again and landed on the glacier surface next to the second huge block. It was partially split, leaving a wide central air gap—"The Keyhole," we immediately christened it. With some manoeuvring, we staged a photograph. (Figs. 38, 39, 40, 41, 42) David is seen through the cleft, steering the helicopter to appear as if he were about to put the "key" (the helicopter) in the "hole." In the fullness of time, the glacier was officially named the Keyhole Glacier, and the high mountain perch at 1,640 metres asl became Shadow Mountain.[7]

The main summer operations of 1966

This year the field operations had begun on May 10. Staff and equipment reached the various early season field sites from Frobisher by seven chartered DC-3 ski-equipped flights between May 10 and May 26. During this early period, the main activities centred on the Barnes Ice Cap. After the middle of June, the various geomorphological groups were airlifted to their respective sites, mainly by helicopter. David Harrison arrived at the Inugsuin base camp with CF-IFF, a Bell G2A similar to CF-FCK, on June 19 and remained in great demand until the very end of the season, on August 28. A Cessna 185, under charter from Gander Aviation (Newfoundland), reached Inugsuin on August 9 and provided vital backup and relief to the helicopter, considering that the total field party numbered twenty-seven. Once again, the breakthrough of women in the field was demonstrated: graduate UBC geomorphologist June Ryder was assisted by Jean Logie, and Jane Philpot of the permanent branch staff was aided by Penny Crompton, a geography undergrad from the University of Toronto. The organizational/staffing arrangements reflected the reorganization of the department—now the Department of Energy, Mines, and Resources. Thus, work on the Decade Glacier had become the responsibility of the Waters Research Branch: specifically, Ken Simpson, aided by regular Geographical Branch summer students.

There were two additional significant extensions. Dr. Rolf Feyling-Hanssen joined us from Denmark to undertake a detailed investigation of the cliff sedimentary stratigraphy along the Clyde Foreland. In September, Olav, supported by the Marine Science Branch (Bedford Institute of Oceanography), secured the co-operation of the DOT: it allocated the Canadian Coast Guard ships *CCGS Labrador* and *CCGS d'Iberville*. Olav was thus able to track seafloor and fiord-bottom topography along the northeast coast over a total distance of close to five thousand kilometres.

Fig. 38: Another high summit: Shadow Mountain, 1,640 m. Tors (right of helicopter), frost-shattered bedrock, and no trace of the former presence of glacier ice. (Photo: August 1966)

Fig. 39: A summit with a view. Looking down onto what was later named "Keyhole Glacier," another textbook example of a glacier with lateral, medial, and multiple end moraines, and also two "small" rocks (shown by arrow) on its surface. (Photo: August 1966)

Fig. 40: Helicopter closeup during the descent to inspect the "small" rocks on the surface of the glacier. (Photo: August 1966)

Fig. 41: The chopper, pictured for scale, gives a better sense of the size of one of the "small" rocks. Recent satellite imagery shows that both boulders have moved down-glacier over the last fifty years and are now hard to distinguish amongst the end moraine material. (Photo: August 1966)

Fig. 42: A very rare example of a chopper stunt. The second giant erratic had a major flaw: the keyhole—hence the name "Keyhole Glacier." David never came close but we felt the need for a simulated fly-through. The first giant erratic can be seen to the left of the helicopter. (Photo: August 1966)

The summer activities also took on an entirely new aspect. We had been contacted by the Canadian Wildlife Service senior animal ecologist, Dr. Doug Pimlott, with a request to assist (insofar as assistance was practical) a visiting *Time-Life* photographer, Stan Wayman. Wayman's large corpus of work had included images of President Lyndon Johnson's family, shots of the first U.S. Beatles concert, and a wide range of nature photographs. His mission for *Life* magazine on this occasion was to photograph denning activities of the Baffin Island subspecies of the grey wolf (*Canis lupus manningi*). As he intended to visit Fox-3 (Dewar Lakes), Wayman would be close to our general area of activities, making the provision of limited logistical assistance manageable. I agreed to assist. It appeared that the Baffin Island project had evolved into a world-class interdisciplinary and international research undertaking, attracting the attention, albeit indirectly, of an internationally renowned magazine.

The Barnes Ice Cap in 1966

Brian Sagar, with summer students Art Froese and Derek Reimer, continued the glacier mass balance studies across the northern half of the ice cap, while Olav, assisted by Dick Cowan, John England, and Peter Lewis, tackled the southern half. The stake network was again expanded. A total of 280 stakes were installed at intervals of approximately 1.25 kilometres. This included three courses of stakes around the southern dome that were also to be used for precise movement survey. To this end, they were surveyed in to base lines off the ice cap. This work was facilitated by P. Henderson of the departmental Geodetic Survey Division. Given the precise survey of the baseline, a resurvey was planned for the following year, the long time gap between surveys necessitated by what was anticipated to be very slow rates of ice movement. (Fig. 43)

Field facilities were greatly improved over previous years (with the large hut at the head of Inugsuin

Fiord, and the growing number of A-frame huts as designed by Gunnar Østrem, it was beginning to look as though we were there to stay indefinitely). A new A-frame hut was set up beside Generator Lake. At the same time, the DC-3 flew in a J-5 Bombardier tractor together with a six-by-ten-foot Wannigan and sleigh. This combination greatly aided mobility on the ice cap. The much-improved living conditions also enabled onsite analysis of the accumulating field data so that anomalous results could be checked instantly, compared with the far less favourable situation of having to attempt sorting out "strange" data the following winter in Ottawa.

The ice cap mass balance for the 1965–1966 budget year proved positive, and the initial 1964 finding of northeast-to-southwest mass balance asymmetry was reconfirmed. Although less pronounced, the same phenomenon was detected for the northern half of the ice cap. Finally, initial examination of the ice cap surface survey indicated that a section of its southern margin had experienced a sudden slump forward at some time in the past. With the work completed, both field groups assembled at Generator Lake on June 14 and, with the later help of the helicopter, were dispersed to field areas off the ice cap now that the land surface snow cover was beginning to melt. Brian returned to Ottawa.

Decade Glacier survey and glacio-hydrology

Ken Simpson, assisted by Tim Sookocheff, continued and intensified the Decade Glacier study that had been set up by Gunnar Østrem in 1965. Measurements of the previous winter's accumulation and the 1966 summer ablation were undertaken from May 24 until August 26. A meteorological station was maintained near the camp at 950 metres with observations at twelve-hour intervals. The by-no-means-unusual problems that faced glaciological research reduced the value of results from the early part

Fig. 43: Southern dome of Barnes Ice Cap. Generator Lake (foreground) is dammed against the ice cap. It spills into Clyde Inlet. If the ice cap were to melt, it would empty its waters in the opposite direction, ultimately into Foxe Basin. (Photo: July 1966)

Fig. 44: The helicopter has just delivered a snowmobile to the camp high on the Decade Glacier. Tim Sookocheff is retrieving the machine while David readies to return to Inugsuin base. (Photo: August 1967)

of the season, especially the high winds and blowing snow that limited the value of precipitation measurements. A detailed topographic survey was made with the precise measurement of a baseline and control stations. This would then provide ground control for the air photography to be attempted in August, the objective being to produce a closely contoured map, scale 1:10 000, of the glacier and surrounding terrain. (Fig. 44)

Bill Rannie installed automatic recording gauges both on the Decade River, close to the level of the fiord, and on the Inugsuin River above the base camp. He collected a large number of water samples in order to calculate sediment yield and chemical composition. While this type of work was still in its early stages, considerable differences between 1965 and 1966 in all the meteorological, glaciological, and hydrological elements relating to the Decade Glacier were recorded.

Home Bay and Ekalugad Fiord

Several small, semi-independent groups began their research in the Home Bay–Cape Hooper region of the northeast coast. John Andrews tackled the moraine systems and their relationship with the raised marine shore features. He concentrated on the distal (Baffin Bay) side of the prominent Ekalugad moraine that had been observed during reconnaissance missions in 1964 and 1965. Jane Philpot studied the inland (proximal) side of the moraine. Both studies were extended to similar situations in adjacent fiords. In the general Home Bay area, the fiords are much smaller and the terrain less rugged than farther north or south, with a result that the depositional features, both glacial and marine, are much more frequent and better preserved. (Figs. 45, 46, 47)

The inner part of Ekalugad Fiord had been dammed by a large end moraine (Ekalugad Moraine) laid down by an outlet glacier that subsequently retreated inland. The lake that formed behind the moraine dam was gradually being partially filled in by glacial outwash gravels. Eventually the lake overflow broke through the moraine dam as sea level continued to fall during the final phases of the last ice age; this resulted in formation of a lower outwash plain (glacial sandur—an Icelandic term) as the river aggraded to the lowering sea level. Mike Church, assisted by Barry Goodison and, at times, several others, began a study of the multi-layered sandur. Conjointly, June Ryder (later Church) studied the formation of talus slopes and alluvial fans in the same vicinity, an experience that figured significantly in the subsequent development of her career as geomorphologist.

In effect, four small semi-independent field parties operated in the Home Bay–Ekalugad Fiord area throughout most of the 1966 summer. These operations depended almost totally on helicopter support until later in the season when the Cessna could find some open water. John Andrews served as the overall party leader and advisor. The combined effort of these four teams led to the construction of six uplift curves as well as the determination of amounts, rates, and direction of isostatic tilt of the land. The significant fiord-to-fiord variation in heights of the marine limit allowed for determination of the relationships between late-glacial rates of retreat of the main outlet glaciers.

Ancillary studies were undertaken, such as collection of vascular plants for identification by Pat Webber, who had not accompanied us on this occasion, and the recording of dates of flowering and seed-set of many species of vascular plants growing in different micro-environments. Jane also extended Cuchlaine King's work of the previous year that entailed calculation of pebble roundness; this was to test the sensitivity of Cuchlaine's method of identifying local changes within different geomorphological environments such as deltas, marine beaches, and alluvial fans.

Fig. 45: First reconnaissance of the Ekalugad sandur (multi-level glacial outwash plain—Icelandic sandur). The Ekalugad end moraine, sandur, and fiord became an intensive study site, especially for Mike Church. (Photo: July 1965)

Fig. 46: From a much lower altitude than in Fig. 45, the two main levels of the Ekalugad sandur are clearly visible. The older, vegetated surface is split by the giant cracks of ice-wedge polygons. (Photo: July 1965)

Fig. 47: The Ekalugad end moraine. Originally the end moraine held back a lake that was forming as the glacier retreated several thousand years ago, when sea level was higher than it is today. As sea level fell and the weight of water in the lake continued to increase, the moraine dam partially collapsed and the lake became an extension of the fiord. (Photo: July 1965)

Clyde Foreland cliffs

Between June 13 and July 27, Dr. Rolf Feyling-Hanssen surveyed cliff stratigraphic profiles along the thirty-kilometre stretch of sea cliffs between Cape Christian and the outlet of the Kogalu River. The substantial and expensive helicopter support, which entailed five camp moves, helicopter time, and fuel between the Inugsuin base camp and Cape Christian, could be given no relief by floatplane support because of unreliable open-water access. Rolf was assisted by Art Froese, Derek Reimer, and Peter Puxley. Twenty-nine profiles were plotted, ranging in height from forty to sixty metres. Large collections of mega-fossils (seashells) and micro-fossils were taken, together with a considerable number of sediment samples. Above the cliffs, the gently sloping surface was also carefully examined. The multiple cliff strata included several layers of glacial till intercalated with marine and fluvio-glacial sediments, comprising the most extensive glacial-interglacial sequence yet discovered in the Canadian Arctic. Rolf believed he had evidence to support the contention that at the height of the last glaciation most of the coastal foreland was submerged by glaciers expanding out from the mouths of Clyde Inlet and Ayr Lake, while sections of land between them remained ice-free. These included the several prominent low hills, such as Tall Sentinel and Small Sentinel, that were believed to have been shaped by ice into moulded forms prior to the last ice age. Analysis of these extensive fossil and sediment collections would take several years.

Remote Lake and the Far Northeast

The peninsula that extends into Baffin Bay between Sam Ford Fiord and Scott Inlet had for several years presented a logistical challenge: it had been impossible to reach it. The outer part was a continuation of the coastal lowland. It was bordered by a series of large lateral moraines that swept out of Sam Ford Fiord,

eventually disappearing beneath sea level as they crossed the coastline. They also dammed a sizable lake, and air photographs showed evidence of different former sea levels intermingled with the moraines. I had regarded the locality as a potentially important one, although its location at the extreme northeastern extremity of our field operations, together with the unlikeliness that the lake ice would break up enough for pontoon landings, constituted a logistical challenge. The total distance from base camp was beyond the capacity of the helicopter. (Figs. 48 and 49)

We solved the issue by finding open water in one of the subsidiary branches of Gibbs Fiord, within Scott Inlet, where we used the Cessna to lay a fuel cache for David and CF-IFF. We were able to work out of a light tent camp between August 7 and August 11. Jane Philpot, with Penny Crompton as assistant, David, and I managed an intensive four days of fieldwork. We mapped the lateral moraines, surveyed the uplands, and collected four marine mollusc samples from different higher sea level stages alternating with several of the lateral moraines. This undertaking also provided a strong link with the submarine ridges extending far across the continental shelf that were later mapped by Olav, from *CCGS d'Iberville*.

An excursion into the realm of the Baffin Island wolf

Later in the season, Stan Wayman contacted us by radio from Fox-3. At the time, we were maintaining a temporary camp some forty kilometres farther east, so it was only a short hop to pick up the mail from Fox-3 and ferry Stan to join us. He spent several nights with us, including an all-night vigil close to a wolf denning site. He eventually obtained, among many other striking images, a fabulous closeup of the she-wolf's eyes, subsequently published in *Life*. He proved a most convivial companion, showing great patience in offering advice on photography. His main "complaint" that we should increase by at least

Fig. 48: Mount Longstaff and Remote Lake. Mount Longstaff (height 5,384 m) is the sharp peak on the skyline, left of centre. It was named after Tom Longstaff, who, as a member of J. M. Wordie's expedition in the 1930s, initiated the earliest mountaineering in northeastern Baffin Island. Prior to that he was a leading member of the first reconnaissance expedition to Mount Everest from the north. This sighting was made during our exploration of the Remote Lake region on the outer coast of Baffin Island. (Photo: August 1966)

Fig. 49: Gibbs Fiord and Remote Lake. View toward the southwest along the length of Gibbs Fiord. The fiord wall reaches heights of 1,200 metres. Depth soundings of more than 1,000 metres make this gigantic feature significantly deeper than Arizona's Grand Canyon. (Photo: August 1966).

an order of magnitude the amount of film we were shooting—which was much more than we had available—although I think he had rather more extensive "home darkroom" resources at his disposal than we could even dream about.

Submarine geomorphology: Baffin Bay continental shelf and eastern fiords

Olav was able to work with *CCGS Labrador* and *CCGS d'Iberville* between September 5 and September 25. During this period, over three thousand nautical miles (as measured by the ships' instruments) of soundings and submarine hydrographic tracks were taken between Sam Ford and McBeth fiords out to the edge of the continental shelf as well as along the length of four of the major fiords. This provided an extensive record of the sea-bottom topography relevant to a fuller interpretation of the land surveys. The fiords proved to be of classical shape (that is, eroded by thick glaciers) with their deepest parts coinciding with the highest mountain summits, giving evidence of considerable glacial overdeepening. This exceeded 900 metres for Sam Ford Fiord. As the surrounding heights rose to 1,700 metres above sea level, total relief was greater than 2,600 metres: a greater depth/ height than the Grand Canyon.

The fiords also shallowed toward their mouths, with thresholds of less than two hundred metres, only to deepen again as great troughs extending across the continental shelf. The shelf varied between forty and sixty kilometres in width with trough depths exceeding eight hundred metres. The troughs themselves were bordered by low linear ridges, in some cases being conspicuous extensions of the lateral moraines that had been mapped on the outer coast. Detailed fiord cross-sections were also obtained using the main vessel's motor launch. On the journey south, all the main fiords were observed to extend as great troughs across the continental shelf, including Ekalugad Fiord and all the way south to Broughton and Padloping islands.

End of the 1966 summer field season

With the exception of Olav's September "at sea," all the field personnel had returned to Ottawa by the last day of August. This sixth of the Baffin Island summer expeditions felt as though it had been a triumph, and Olav's subsequent deep-sea excursion was more than icing on the cake. The 1966 expedition rounded off a level of interdepartmental, international, and interuniversity cooperation that never could have been envisaged in 1960. It also provided a record of solid field research and outstanding rapport among permanent staff, senior invited colleagues, students, aircrew, and DEW Line personnel. The relations with our renamed department's new Waters Research Branch, which had originally caused me so much concern, were excellent; indeed, work in the field had proceeded as if no administrative restructuring had occurred. Above all, my sense of responsibility for so many exuberant students, male and female, while considerable, had proved exhilarating. There had been one or two inadvertent cold baths, but not even a sprained ankle had come to the attention of Olav or myself.

Fig. 50: A precarious helicopter perch, inner Clyde Inlet. One small ledge (shown by arrow) surrounded by glaciers and rock faces was sufficient for a helicopter perch to enable collection of rock and soil samples. (Photo: July 1966)

CHAPTER 9

⚡ LAST YEAR OF BAFFIN ISLAND ⚡ ACTIVITIES BY THE GEOGRAPHICAL BRANCH, 1966–1967

The most successful Baffin Island season to date—that of 1966—was augmented by advances in many other areas in Canadian government geography. The newly designed desk-size *Atlas of Canada*[1] was beginning to take shape. A study had been completed on the impact of tourism on the economy of Prince Edward Island; it included a map of the extent of shoreline loss to public access due to land purchases for private summer cottages, especially by the rich of New England. A regional study of the Cypress Hills was in its second season. The Minister of Transport had strongly approved the first phase of investigation of the Prairie railway branch line system by Tony Burges. Professor Alice Coleman (University of London), director of the Land Use Survey of Britain, whom I had tempted to join us during her sabbatical leave, was well into a collaboration with the National Capital Commission and the City of Ottawa in a study of city, greenbelt, and surrounding land-use and land-planning problems. There were many other geographical activities.

The third meeting of the new national advisory committee on geographical research had been held in May 1966, with a second round of research grants awarded to academic geographers. And I was being invited to give lectures—on both the Baffin Island program and the general work by the branch—by university departments of geography across the country as well as to local service clubs. So I remained confident that my plans for the branch would meet with success.

Major changes afoot in the department

The department—now the Department of Energy, Mines, and Resources, as of June 1966—was changing rapidly, however. Dr. van Steenburgh had officially retired but was retained for a year in a science advisory capacity. The new DM, Dr. Claude Isbister, was an unknown quantity to me. Nevertheless, I was hopeful because of slight contacts through our mutual membership in the Ottawa Unitarian Congregation, our church, although I perceived him as a somewhat remote personality. Below the deputy minister level, there were four assistant deputy ministers (ADMs) to oversee a much enlarged departmental structure. Dr. Jim Harrison, as ADM for Earth Sciences, remained as my immediate superior. The Inland Waters Branch was now part of a much larger structure, with its own ADM whose responsibilities included the newly formed Bedford Institute of Oceanography and a Great Lakes water research institution. This placed glaciology beyond the immediate purview of "earth sciences," let alone the Geographical Branch.

On the positive side, relations with the growing number of collaborating federal institutions and the universities were excellent. Then, in early December, Dr. Van invited me into his office for a chat. He presented me with a surprise report: the *Organizational Study of the Department of Energy, Mines, and Resources* was dated July 1966, but with a notation to the effect that it be made available to a restricted readership only in December 1966. This report was purported to have been based, in part, on interviews with all departmental senior officers, including branch directors. Dr. Van smoothed over my objection that I had not been interviewed by reminding me that there had been three meetings of the National Advisory Committee on Geographical Research as well as the extensive formal discussions of April 15 and 16 that he had chaired and for which I had prepared the very extensive minutes. He pointed out that the new report had not proposed any move or changes for the Geographical Branch. Nevertheless, it contained a somewhat alarming, for me at least, "Recommendation 19," which read as follows:

> Careful consideration be given as to the proper role of the Geographical Branch and a determination made as to whether or not it is really a support function and whether its divisions would be more appropriately located in various branches of the department.

Dr. Van then produced for me a copy of the much larger Part 2 of the report and went over the more specific statements relating to the Geographical Branch.

> It is not possible on the strength of the investigation . . . to make any firm proposals as to the proper role and organizational location of the divisions of the Geographical Branch. Such proposals would require detailed study of the functions of this Branch. For the present, it has been recommended that the Branch be

located under the Assistant Deputy Minister, Earth Sciences. It is, however, quite evident that some serious thought should be given to this problem.

There followed a longish paragraph that discussed the branch divisions of Economic Geography, Physical Geography, and Toponymy, concluding with an encouraging statement

> that careful thought should be given to the problem and a decision made which will be in the best interests of both the Department and the staff of the Geographical Branch.

Dr. Van urged me to discuss Recommendation 19 with Jim Harrison, pointing out that Jim was a personal friend, not an adversary, who held me in considerable esteem. He assured me that I should feel free to reason with Jim, but, as a partially retired advisor, he (Dr. Van) had no formal authority. He thought that any final decision would likely be made at a much higher level. He urged that there was no need for me to worry, that the very significant progress of the branch had been clearly recognized. I was to be given a full promotional step, retroactive to the beginning of 1966.

In this manner, the position of director of the Geographical Branch was placed on the same level as that of the Geological Survey. Furthermore—and much more important to me, as it had been the issue of a long personal struggle—the "geographer" professional service steps were to be upgraded, in terms of salary brackets, to match those of all other disciplines in the federal civil service. Dr. Van advised me not to discuss the content of the organizational report, but told me that I should feel free to relate the obviously morale-boosting news about salary levels. He also thought I could inform members of the National Advisory Committee about the salary scales. Nevertheless, from all this it was apparent that I was due for a hard winter, although Dr. Van advised that I should remain optimistic and do all I could to shield my staff from any unnecessary concern—or, as I read him, keep the difficulties largely to myself.

This critical meeting was followed up by a most convivial one with Jim Harrison. Jim also assured me that I should not worry but confessed that he could not resist reiterating his earlier offer for me to serve as head of a large new Division of Quaternary Research within the Geological Survey. Again I refused, but this time I acknowledged his considerable compliment. He assured me that my personal future prospects remained very bright.

A surprising invitation

Then, in an almost surreal circumstance, early in the new year I received a startling telephone call from a Jim Archer, Graduate Dean at the University of Colorado, Boulder. He was on a personal mission to recruit a new director for the university's Institute of Arctic and Alpine Research—and I was his preferred candidate. Would I accept an invitation to visit Boulder? I explained that I was a Canadian by choice (that is, an immigrant), that I already had a most challenging position, and that I had no wish to leave Canada. However, I allowed him to persuade me to accept an invitation to give a lecture on the Baffin Island project and spend three days on the Boulder campus. When he said that the invitation was to include my entire family, I replied that I would rather come alone as I did not want to create a sense of personal obligation.

I left Ottawa for Denver on February 13, 1967. The morning temperature at Ottawa's airport was –30.6°C (–23°F). The next morning, Boulder was enjoying bright sun, its impressive mountain backdrop, and a temperature that was about 26°C (80°F) higher than that in Ottawa. For three days I was treated royally—group dinners; a day's visit to the Mountain Research Station in the Front Range, at three thousand metres; and an overwhelming response from a large audience following my lecture. I was finding it very hard to say no to Archer's job offer. So I devised a response that would make me totally unaffordable: a demand for six new faculty positions, support staff, financial support for a new quarterly scientific journal (complete with editor), and a new building. To my acute embarrassment, all of my "conditions" were accepted as reasonable. But why me? I was told I had received exceptional recommendations from Dr. A. Lincoln Washburn and Professor J. Ross Mackay, senior colleagues and advisors.

Dean Archer personally drove me to the airport the final morning (as, indeed, he had met me on arrival), although I left him with a firm refusal. But as the following weeks slipped by, I began to rethink the situation in view of certain dark clouds that had begun to gather on the horizon. Two things happened, one minor, one major, that proved influential. But so far as I could see, they were not connected.

One morning toward the end of February, Bob Code, director of personnel, telephoned to say that the Ives family had been accorded a considerable honour. We had been selected as one of the first of twelve families in a new federal government program for a full year's total immersion in Québecois culture. He knew that I would receive a strong welcome from Université Laval and this would be an excellent opportunity for all of us to become proficient in French, which in turn would greatly improve my prospects for advancement in the civil service.

The second was a late-March telephone call from the deputy minister, Dr. Isbister. He calmly informed me that the Geographical Branch was to be disbanded and that I should begin to consider my personal future. I was aghast. This, for me, came completely out of the blue. I protested, only to be met by a gruff statement that there was to be no discussion. He urged me to take a week to think it through, and then to arrange for a meeting with him—but to keep my decision strictly confidential.

After six days, I telephoned the DM's secretary and asked for an interview. I was told to try again the next day as he was too busy. I tried for three days and was rebuffed at each call. On the final try, I asked the secretary to inform Dr. Isbister that unless he responded, I would call a full staff meeting that afternoon and deliver the bad news. That quickly brought him to the phone. He reminded me that the decision was strictly confidential and would remain so until the following August. I explained that he had put me in an impossible situation. I felt a strong obligation to my staff, and I was sure that several of them would want to apply for university positions. I was amazed to hear him say that was why he wanted the decision kept confidential: he didn't want to lose research staff to the universities—at least, not until so late in the year that anyone who wanted to apply would be too late. Upon my protest that the situation was morally intolerable, he asked if I knew the meaning of the word "insubordination." This both alarmed and angered me. I found myself trembling but told him to send me an instruction in writing and I would give him the pleasure of flouting it. I said I would go ahead with a full staff meeting that afternoon. The last words that I ever received from Dr. Isbister were these: "Ives, as far as I am concerned, you no longer exist." This was completely beyond anything I could have expected. Was I at risk of being fired? I called in my secretary, Donna, and asked her to set up a full staff meeting for that afternoon.

A major reorganization—and the response

The afternoon was decidedly tense. Everyone was shocked, although I assured them that nobody was in danger of losing their jobs—that it was "simply" a question of departmental organization. I also reminded them that, after a long struggle, the formal "geographer" job classification had been upgraded to match that of the Geologist scale and every other professional scale in the civil service. I admitted that in calling the meeting I had defied the specific

instructions of the DM, and that I would explain the reasons for this privately to anyone who might wish to know. Understandably, I had several requests for private interviews.

After the meeting, I worked with Donna to prepare telegrams, with longer follow-up letters, to send to the chairs of all Canadian departments of geography and to members of the National Advisory Committee, whose volunteer service over the past two years I judged to have been mocked. I added that neither I nor, to my knowledge, any single geographer had been consulted about the decision to dismember the branch. Copies were delivered by hand to Dr. Isbister and Dr. Harrison.

Within several days there was an immediate response from across the country. The minister, the Honourable Jean-Luc Pépin, was deluged with telegrams and phone calls. Many demanded a full explanation, others an urgent reconsideration. All condemned the decision and especially the way in which it had been made. Dr. Harrison asked me to his office. I was very nervous; however, my feelings were immediately relieved as he sat me down and cordially poured coffee. Then he made an inevitable remark, to the effect that I had certainly created a stir. My breach of confidentiality on such a large scale had caused an enormous reaction. The minister was extremely upset; Dr. Isbister was outraged. Dr. Van, at least, and as usual, had provided some light relief (for Jim Harrison, that is—not Isbister). Jim said that during the immediate high-level conclave in the DM's office, Dr. Van had half-laughingly reminded the others that he had warned this was not the way "to handle Ives." Nevertheless, Minister Pépin had telephoned Jim and asked him to let me know that he, the minister, wanted to see me in his office on Parliament Hill. Pépin had explained that his secretary would notify me directly of a time and date.

I couldn't help asking Jim whether my portending visit to the minister's office would take place in the formal custody of a red-coated officer. He replied that it was likely the minister would try to smooth things over for me and ask for suggestions on how to mollify the Canadian geographical profession. Jim explained that I certainly shouldn't worry too much, although I should not expect any further contact with the DM. Jim saw me out with a warm handshake, a real act of friendship, and sent me back to my office, where I then tried to unravel what was happening.

Pépin had been appointed Minister of Energy, Mines, and Resources only a few months earlier. At the time, I had written him a formal letter of congratulation; it had included information about the branch's adoption of Samuel de Champlain as our Canadian geographical hero and about our use of Champlain's famous astrolabe as the model for both our new branch shield and the cover of the *Geographical Bulletin*—a copy of which I had enclosed. Pépin's response had been rapid and extremely warm, applauding our adoption of Champlain and concluding with the following line:[2]

> Your new symbol represents a challenge to all members of the Geographical Branch—to follow in Champlain's footsteps is not an easy task.
>
> —Jean-Luc Pépin, 16 January 1967

Within a week of my meeting with Jim Harrison, my nervousness had by no means abated as I approached the minister's office. I should not have worried. I was received agreeably, and I found myself in the presence of a singularly attractive person of distinguished bearing. There was a choice of tea or coffee, together with a plate of cookies. The minister expressed his personal regrets, while at the same time indicating that I had certainly caused a great commotion and a lot of extra work for him. He then went on to say that he hoped I would help him to come up with ideas on how to appease the universities. He also explained that there was no way in which the decision could be reversed and that it had been made prior to his appointment as minister.

There followed a relaxed conversation in which the minister posed numerous questions concerning my views about geography and Canada, but especially about what he could do to help close the rift that had been created. After some discussion, he agreed to intervene with the upper administration of the department to ensure that $30,000 would be budgeted to enable the 1972 International Geographical Congress to be held in Montreal.[3] Next, he would do all he could to support the National Atlas program and the *Geographical Bulletin*. He explained that Dr. Harrison was enthusiastic about the Baffin Island project and that its continuation was assured. Finally, he accepted my request that he offer to make a speech during the annual meeting of the Canadian Association of Geographers. Conveniently, the 1967 meeting was scheduled to be held at Carleton University (in Ottawa) within a few weeks. In order to prepare and to say the "right things," the minister asked if I would write a draft speech for him. As I was taking my leave, he wished me a good summer in Baffin Island; he then expressed surprise and disappointment when, in thanking him, I explained that, with regret, after my final visit to my Arctic paradise, I would be moving with my family to Boulder, Colorado.

That was the only time I met with the Honourable Jean-Luc Pépin. A few days after that personal meeting, he sent me the final text he intended to present to the Canadian Association of Geographers. I was elated to see that he was accepting practically all of my suggestions as his commitment in writing. And he had added a variation on one of my favourite quotations for his concluding remark, thereby identifying the source of at least some of his commitments, as my enthusiasm for this quotation was well known:

> As Mrs. Malaprop says in Sheridan's *The Rivals*, "I would have her instructed in geography that she may know something of the contagious countries."[4]

From Baffin Island to the Colorado Rockies, 1967

The 1967 field season proved an extremely strange experience for me. First, considering the logistics, the total number of personnel, and the number of contributing institutions, it was the most ambitious and complicated operation of the entire seven-year Baffin period. The air support included the usual early season DC-3 on ski-wheels, with the first group of staff reaching the field at the beginning of May. On this occasion, there were two Spartan Air Services Bell helicopters, the second chartered jointly with the Canadian Wildlife Service, which had a team camping in the Nettilling and Amadjuak lakes area on a study of the summer nesting ground of the blue goose. David Harrison was the chief helicopter pilot and Jim Crawford the second. (Fig. 51)

Our late request for a second helicopter, however, resulted in an unusual turn of events. Spartan didn't have one available; it was necessary for the company to purchase a brand new Bell 47 G4A (with a more powerful engine). David had to fly down to Fort Worth, Texas, in May to pick it up and fly it back to Ottawa. As the Canadian registration did not come through in time, David operated all summer under the U.S.-registered N7919S. Tom Murray, our prized engineer and baker of fresh bread, was not available this summer. Jim O'Shaughnessy ably took his place and serviced both machines from the Inugsuin base—but didn't bake us fresh bread for dinner.

A Cessna floatplane joined us for the later open-water period. (Fig. 52) As in 1966, *CCGS d'Iberville* was made available in September for more extensive submarine mapping along the northeast coast. Mike Church pushed the field season to its extreme limits by continuing his work on the Ekalugad sandur until he was picked up by icebreaker in October.[5]

The federal institutions collaborating in Baffin Island in 1967 included the Inland Waters Branch, the Bedford Institute of Oceanography, the Canadian Hydrographic Service, the Surveys and Mapping Branch (for extensive air photography leading to a series of excellent 1:50 000 topographic maps), and the Geodetic Survey for radio-echo sounding and strain rate determinations on the Barnes Ice Cap.

All these activities related to the core of the Geographical Branch operations. Contributing universities included (and university personnel came from) Queen's, McGill, York University, and University of Manitoba, as well as Nottingham and Southampton, in the U.K., and Yale, Maine, and Wisconsin, in the United States. Thus, many academic disciplines and technical surveys were accommodated under the general branch leadership; hence the personal sense of strangeness that I am sure was felt by many of the participants. The fate of the branch had been sealed, and it would cease to exist as an entity before the end of 1967. After my departure for Colorado, Dr. Keith Fraser undertook the onerous duties of acting director. While Canadian government research activities on Baffin Island were to continue for several more years, many of the principals left Ottawa. Even a number of Canadian students chose to follow suit, several completing their postgraduate degrees at the University of Colorado. The total number of participants in the 1967 season was logged at thirty-eight, of whom five were women.

The DC-3 flew north from Ottawa on May 10 with Martin Barnett and Doug Christian, our excellent first full-time resident base camp manager, who were in charge of the supply flights. Olav Løken, along with his Barnes Ice Cap group, worked intensively across the south dome between mid-May and the end of June. Most of the others were in position with helicopter support by the end of June. Olav returned to Ottawa in late June and I did not arrive until July 21. John Andrews operated very much "off-centre" and unusually late in the season. He spent the period from September 11 to September 22 making a reconnaissance of Broughton Island, some 150 kilometres south of the main field area. This was the first phase of what had been intended as a major extension of the research activities on southeastern

Fig. 51: Inugsuin base camp in 1967. The double helicopter charter of 1967 greatly intensified the fieldwork, easing the pressure to move personnel and equipment to distant field camps. (Photo: August 1967)

Fig. 52: The Cessna proved a great support during the open-water season, and where landing possibilities could be located, it significantly reduced the cost of extensive dependency on the choppers. (Photo: August 1967)

Baffin Island.[6] Thus, for the period from late June until my arrival on July 21, the expedition ran itself. The operation went remarkably smoothly. While many of the personnel were Baffin veterans, Doug Christian and David Harrison played vital roles by maintaining a gentle control over the many camp moves that were needed.

Glaciology and hydrology

The 1967 season included studies of ice thickness, surface movement, and mass balance of the south dome of the Barnes Ice Cap, undertaken by Olav and several assistants. Maximum thickness at the south dome was 550 metres, while the underlying bedrock surface varied between about 300 and 550 metres asl. This confirmed the work of previous years farther north that showed the ice cap sits on an undulating plateau with surface characteristics comparable to those of the surrounding ice-free areas. Surface movement was determined as very slight, ranging between 2.29 and 24.19 metres in the course of the previous twelve-month period. Again, there was further confirmation of the asymmetric nature of the mass balance, with higher accumulation along the northeast side of the ice cap. Strain rates were also determined along all three margins, and evidence was obtained that implied some form of glacier surge had affected the southwestern margin in the past.

Work on the Decade Glacier, under the direction of Alan Stanley of the Inland Waters Branch, repeated that of 1966 and recorded another year of negative mass balance. Hydrographic and climate stations were maintained on the lower Decade and Inugsuin rivers, as well as at the climate station high on the glacier.

Roger Barry spent the period from June 26 to July 16 working between Inugsuin base camp and the Barnes Ice Cap. He established three unmanned climate stations, one on the crest and one each beyond the southwestern and northeastern margins. He also undertook to improve the instrumentation and recording capabilities of the short-term climate stations at base camp, Ekalugad Fiord, and other locations. The anticipated data were intended to feed into a synoptic climatology of Baffin Island. Roger was on sabbatical leave for the year from the University of Southampton, U.K.[7]

Glacial geomorphology

This season saw a total concentration on the Home Bay area of the northeast coast. Cuchlaine King, assisted by Lyn (Drapier) Arsenault and Mary (Strom) Millest, worked from seven campsites extending from the high mountains of the midsection of Henry Kater Peninsula to the outer coast. Their work involved tracing the multiple moraine systems and surveying raised marine beaches and the marine limit;

they also collected numerous mollusc shells for radiocarbon dating.

John England, assisted by Tim Sookocheff and Patrick McLaren, undertook similar studies throughout the Home Bay fiords to the south, extending the work begun the previous year by John Andrews and Jane Philpot. Together the combined work of the two seasons would provide extensive data for mapping the profile of the deglacial marine limit and drawing isobases (contours) on lower emergent shorelines throughout the Home Bay area (the shorelines were tilted up toward the southwest). Efforts were made to match different sea levels and late-glacial retreat phases that showed the pattern of withdrawal of the outlet glaciers onto the central Baffin plateau.

Mike Church, with four assistants, extended the 1966 study of the Ekalugad sandur (glacial outwash plain). With support from Surveys and Mapping Branch staff, a detailed topographic map was made of the sandur and surrounding valleys. Mike's work included detailed studies of river discharge, silt content and sedimentation, pebble counts from numerous localities, and the maintenance of a climate station. As an example of the data-intensive nature of the work, 148 points across the "old" and recent sandur surfaces involved over 46,000 measurements on more than 14,000 sandur cobbles. Mike's research eventually led to his doctoral degree in UBC's department of geography (Church, 1970). A version was also published as a monograph by the GSC (Church, 1972). (Figs. 53 and 54)

June Ryder, assisted by Penny Crompton, continued her work of the previous year with minute survey and measurement of talus slopes and alluvial fans throughout the Ekalugad area. June and Penny also joined with Mike's group during periods of intensive water and sediment sampling.

Fig. 53: The challenge of collecting samples from a fast-flowing river in Baffin Island. A homemade breeches buoy. Mike Church is being pulled across the Ekalugad River by June Ryder and Penny Crompton. (Photo: August 1967)

Fig. 54: The Ekalugad Camp in 1967. Mike Church, in background. (Photo: August 1967)

Submarine survey

Doug Hodgson worked from *CCGS d'Iberville* in September to greatly expand the submarine mapping that Olav had initiated in 1966. Between Pond Inlet in the farthest north and Broughton Island in the south, 5,700 kilometres of survey was completed. This included 3,200 kilometres offshore and 2,500 kilometres along the fiords. The greatest depths were recorded for the Scott Inlet trough, which demonstrated more than seven hundred metres of glacial overdeepening.

Biology

Pat Webber, with three assistants—John Richardson, Bill Phillips, and Ron Irvine—undertook an intensive plant survey in the Home Bay area during the period of June 30 to August 24. Along with his earlier seasons' research, this completed an extensive survey of north-central Baffin Island botany, in addition to the lichenological investigations. I enjoyed a prolonged field visit during which Pat demonstrated his immaculate approach to recording the lichens and vascular plants. It was his recollection, he said many years later, that it was during this field visit that I persuaded him to join me in Boulder, Colorado (P. J. Webber, personal communication, June 2013).[8]

Another aspect of Baffin Island botany was taken up by J. A. Parmelee of the Plant Research Institute, Canadian Department of Agriculture. He was able to visit many of the DEW Line sites and our base camp on Inugsuin Fiord, completing a survey and extensive collection of the fungi of this hitherto mycologically virtually unknown area of central Baffin Island. Parmelee collected over 500 specimens of phaneogams and cryptogams along with soil samples for laboratory identification of soil fungi; he also collected 450 specimens of mosses. The mycological specimens were to be deposited in the Canadian National Mycological Herbarium, and many duplicate specimens were to be distributed as part of the normal herbarium exchange program. The vascular specimens were for deposit in the Phanerogamic Herbarium; both herbaria are housed at the Central Experimental Farm operated by Agriculture and Agri-Food Canada.

The nunatak hypothesis and mountain photography

The 1967 expedition provided my final opportunity to examine mountaintops for evidence of the former presence of ice age glacial activity and to extend my catalogue of landscape and glacier photography. The generally excellent clear weather of the previous two seasons was not repeated, although David, as usual, proved able to grasp every opportunity, especially with a more powerful Bell 47 G4A machine. Especially worthwhile flights were made along Ayr Lake, Clyde Inlet, and McBeth Fiord with repeated summit landings. One particularly unsuccessful operation involved an attempt to re-examine the outer coast immediately south of Clyde Inlet. As we approached the coast we ran into severe turbulence, which forced us to abort any further flight into that area and take refuge on the shore of tiny Bute Island, just off the outermost southern cape of Clyde Inlet. Conditions kept us pinned down there for over twenty hours. On July 16, we were able to escape and struggle our way to the Clyde River settlement. As a result, we had a rare opportunity to photograph several of the local people and purchase a large number of Ookpiks (the entire stock, in fact) from the Hudson's Bay Company store.[9] Then began a very dicey flight back to base camp. The weather remained ugly and very dark, and we had not been able to acquire a satisfactory supply of helicopter fuel at Clyde. But two ten-gallon drums of fuel had been cached just beyond the halfway point. Given the still turbulent weather and the riskiness of even one extra landing, however, once we got to the cache David preferred to just keeping going, having calculated that we could get to base camp on the remaining gas. (Figs. 55 and 56)

Fig. 55: Fog forming over Inugsuin Fiord. Rapid development of fog over the fiord water was a constant threat to helicopter and floatplane operations. (Photo: August 1967)

Fig. 56: Rapid development of the fog reaches serious proportions. (Photo: August 1967)

Last 1967 field engagements

August 19 provided a delightful surprise to everyone present at the Inugsuin base camp—with only a week to go before the season's end, it was a very exciting day. The Canadian Coast Guard ship *Wolfe* suddenly appeared round the corner of the fiord. No one had been aware that it was coming, and all stood agog on the beach as the ship slowly drew near and anchored well away from the inshore shallows. David took the occasion to broaden his helicopter tactical skills with his own first-ever landing on the deck of a ship. He then siezed the opportunity to ferry several of our group over to the *Wolfe*, our women students creating considerable interest among the ship's crew as their presence was a complete surprise. There followed the unloading of fuel and supplies for the following season.

My own final month on Baffin Island had been largely taken up with visiting the field camps, taking additional photographs, and working closely with Doug Christian and David to attempt to satisfy all the understandably heavy demands for helicopter support. Once again, the entire summer operation worked smoothly and was accident-free. All personnel were back in Ottawa by the end of August, except for Mike Church, who was determined to extract as complete a sediment discharge record as possible, and Doug Hodgson, who cruised home with the *d'Iberville*, having recovered Mike and his assistant from Ekalugad Fiord on the way south. (Fig. 57)

My farewell to Baffin Island and (for now) Ottawa

I returned home to a few very hectic days. Pauline had already completed the major share of preparing all of our personal effects to be transported to Colorado and coaching our three small, anxious children for their decidedly unwilling departure from their Ottawa home. My last formal actions were to bid a very sad farewell to the Geographical Branch staff and to visit Dr. van Steenburgh. He wished me well, expressed his regrets that I was leaving, and cheerfully projected that it would not be too long before I returned. Little did either of us imagine that it would prove to be thirty years. I was never to see this wonderful gentleman, scientist, and administrator again.

My final official visit was to my ardent sparring partner, assistant deputy minister Jim Harrison. Jim also expressed his regrets and astonishment that I was prepared to leave Ottawa and the department—how could I take such a course when a great career with the federal government was open to me? I couldn't provide him with an adequate explanation, although I knew in my heart that I couldn't accept the loss of the Geographical Branch. But I did take the opportunity to make a final request: Could I take the Geographical Branch shield with me as a personal memento? Jim readily agreed that I could but cautioned that the shield was not his to give; technically, it was the property of the Department of Public Works. Yet he thought it appropriate that I should have it and, with a familiar provocative grin, advised that I take it down off the wall overlooking Booth Street early one Sunday morning when there would most likely be no one about. He promised to post bail for me if I was caught by the authorities! We shook hands and he wished me good luck. His parting shot was that he hoped to see me back in Ottawa before many years had passed. It was not long thereafter that, despite the difficult administrative struggle, or perhaps because of it, I found that over the years I had earned a most influential friend.[10] I only regret that he did not live long enough to review this manuscript.

Early the following Sunday morning, George Falconer and I raised a ladder against the wall on Booth Street. George held the ladder to keep it from slipping, and I climbed, spanner in hand. My feet were on the second rung from the top before I could reach the top bolt holding the shield to the wall. I worked loose the bolts. As the shield came free, I

Fig. 57: Twilight of the Geographical Branch Baffin Island expeditions. Goodbye to the DEW Line. (Photo: August 1967)

Fig. 58:
The shield of the former Geographical Branch is returned to the staff of a branch remnant—the National Atlas group, Surveys and Mapping Branch, Department of Energy, Mines, and Resources. (Photo: June 2003)

realized that I had not thought through what I was attempting to do. It weighed much more than I had realized and was bulky into the bargain. I swayed dangerously, clutching the shield to my chest, and almost fell backward off the ladder, dropping the spanner, which narrowly missed George's head. The street was deserted. I inched my way down the ladder clasping my precious prize. It travelled with me to Boulder, Colorado, then in 1989 to the University of California, and finally back to Ottawa in 1997. It was to my great pleasure in September 2008 that the *Atlas of Canada* team, from the Surveys and Mapping Branch, welcomed me into their midst and allowed me to "repatriate" the shield to the people of Canada. At last, my long criminal career was at an end. (Fig. 58)

SELECTION
OF ARCTIC
VASCULAR PLANTS,
BAFFIN ISLAND

Fig. 59: *Papaver labradoricum* (Iceland poppy)

Fig. 60: *Dryas integrifolia* (mountain avens)

Fig. 61: *Cassiope tetragona* (Arctic mountain heather)

Fig. 62: *Epilobium angustifolium* (fireweed)

Fig. 63: *Pedicularis sudetica* (Sudeten lousewort)

Fig. 64: *Pyrola grandiflora* (Arctic wintergreen)

Fig. 65: *Salix arctica* (Arctic willow)

CHAPTER 10

CAREER DEVELOPMENT OF
MEMBERS OF THE FIELD TEAM

It was claimed in the opening pages of this account that the subsequent career development of so many of the field team is probably the most important contribution of those seven years of wide-ranging operations and ardent struggle for scientific results. Consequently, this chapter is devoted to brief overviews of the careers of as many former student field assistants and staff as are available to me, given the long lapse of nearly a half century.

A total of seventeen "mini-biographies" are included. Many of those individuals have remained in close, or intermittent, personal contact with me over the years. Others I was able to contact via the Internet, their remarkable success rendering the effort easy.

Not all of the more than fifty team members followed scientific or academic careers. Indeed, one of them contacted me out of the blue a couple of years ago (the first communication since 1967); he wanted to let me know, now he had retired, that while the Baffin experience did not directly shape his career path, it had made a profound and highly beneficial impact on his life. This would indicate that this chapter is somewhat incomplete, as there are others whom I failed to contact.

Another, related aspect is beyond my remit: the impact that these individuals have had on their own students and/or staff under their administrative guidance. As Professor Emeritus John England, now in very active "retirement," remarked in a recent email, we are witnessing the development of a third generation of academics and scientists as well as their influence, which is already extending to yet a fourth generation.

Many have written their own mini-biographies. Few modifications have been made, and the factual information and the spirit of the communication carefully preserved. Most have been checked and approved by their originators. And to preserve all sense of spontaneity, no uniformity of presentation has been attempted.

In thinking back over the last half century, it must be borne in mind that the entire approach to field research, even in regions still considered extremely remote, has changed. The extensive development of technology, instrumentation, and Internet contact, as well as communication in the field, has driven this change. Much of the Baffin Island experience, despite the early acquisition of helicopter support, depended on hiking boots. Chapter 11 expands on this aspect.

Student Field Assistants

MICHAEL CHURCH

Mike was born in Preston, Lancashire, in 1942 and emigrated to Canada with his family at the age of five. He first joined the Geographical Branch Baffin Island project in the summer of 1962 as field assistant to George Falconer in a reconnaissance of the Pilik River basin in northern Baffin Island. Travel was done partly by kayak and, as Mike was subsequently told, he attained the position because he was the smallest, lightest applicant—well suited to kayak travel. He subsequently spent three summers at Lewis River, on the northern margin of the Barnes Ice Cap, monitoring glacial runoff. The equipment provided in the first summer proved inadequate for accurate gauging, so—armed with a copy of Åke Sundborg's classic paper on the River Klarälven in Sweden—Mike decided to study the hydraulics and sedimentology of the outwash plain channels at the glacier front. This experience inspired Mike's subsequent career as a fluvial geomorphologist. Following the Lewis work, he embarked with branch support on three seasons at Ekalugad Fiord, in Home Bay, where his study of the outwash plain became the topic of his PhD thesis, "Baffin Island Sandar."

After these experiences Mike looked forward to a career with the Geological Survey of Canada (which had absorbed much of the Geographical Branch) but was told that his style of process-oriented work was not a priority for the organization. So he remained at his graduate university (the University of British Columbia) initially as a computer programmer, then as a lecturer and ultimately professor. His teaching career, in hydrology, geomorphology, field training, and environment and resource studies, spanned thirty-eight years. His Baffin experience was put to good use in consultancy on prospective routes via the Mackenzie Valley for an Alaskan oil pipeline (1971–1972), further work on Baffin in 1972 and 1976, and work

on the western Arctic coast and Mackenzie River during the 1970s. However, the increasing difficulty of gaining permission for independent work in the Arctic, and the costs, led him to switch regional focus to the western mountains; there, he worked on steep forest streams, contributing to the evolution of forest practice rules for forest land use around streams and to major projects on Peace River and Fraser River. Inspired by Sundborg's example, he is one of the relatively few geomorphologists (until very recently) to tackle research on "big" rivers.

Peace River provided the opportunity to document the response of the river to damming in 1967 for hydropower production, while Fraser River presents the problem of aggradation of gravel in a reach that is adjacent to Lower Mainland settlements but, at the same time, supports one of the richest river fisheries in the world. Mike's focus has been on basic research, most of which has been initiated in response to practical problems and all of it feeding directly into river management issues.

Mike was presented with the Kirk Bryan Award by the Geological Society of America in 1977. In recognition of his river work, he was made a fellow in the Royal Society of Canada. Subsequently, he was elected RSC director of its Earth, Ocean, and Atmospheric Sciences (EOAS) Division of Academy III. He was also elected to fellowship in the American Association for the Advancement of Science, the Geological Society of America, and the British Geomorphological Society (BGS). The BGS awarded him its highest honour, the Linton Award, in 1996. In addition, Mike has been awarded the Massey Medal of the Royal Canadian Geographical Society and the G. K. Warren Award of the (U.S.) National Academy of Sciences. In 2008, Durham University (U.K.) conferred on him an Honorary Doctorate of Science. During the ceremony in Durham Cathedral, he was acknowledged as one of the world's foremost geomorphologists.

[Author's note: As Mike couldn't resist saying when I had to press him for his statement, "Not so bad for the smallest, lightest applicant to go north with the Geographical Branch."]

W. RICHARD (DICK) COWAN

Dick was born in 1942 and grew up on a drumlin south of Ottawa whose shoreline features can be attributed to the postglacial Champlain Sea; it was here he learned some of the realities of boulder till and Leda clay. While attending Carleton University he met John Andrews, who recruited him in 1963 to work on geomorphological studies around the Barnes Ice Cap and in the Bruce Mountains of Baffin Island; he also assisted Olav Løken with the spring accumulation survey on the Barnes Ice Cap in 1966. Dick graduated from Carleton in 1964 (geography) and then accepted an appointment to the McGill Subarctic Research Laboratory in Schefferville (1964–1965), where his research on ribbed moraines led to an MSc from McGill in 1967. He later followed his Baffin/McGill colleagues to the Institute of Arctic and Alpine Research in Boulder, Colorado, headed by Jack Ives, where he completed his PhD in 1975 under the supervision of John Andrews.

Dick's early professional career included ten years of Pleistocene stratigraphy and mapping for the Ontario Geological Survey, where he became supervisor of Quaternary Geology. He then migrated to environmental management and took on technical and management positions with the Northern Pipeline Agency, created to regulate the Alaska Highway Gas Pipeline. After six years of consulting work in Calgary on energy and mining development projects, Dick returned to Ontario, where he took on progressively more responsible management positions with the Ontario government in mineral titles, mineral development, and mine reclamation, eventually retiring in 2005 as director of mines for Ontario. He has published more than fifty papers, reports, and maps on Quaternary geology as well as numerous papers and presentations on mine development, rehabilitation, and regulation. He attributes much of his professional success to the Baffin Island experience, which introduced him to the importance of "good science, sound project planning, and good data."

JOHN ENGLAND

John's first Baffin Island summer experience was in 1965 when he celebrated his nineteenth birthday on arrival at the Inugsuin Fiord base camp. He also participated in 1966 and 1967, followed by completion of his master's and doctoral degrees with INSTAAR, University of Colorado (in 1969 and 1974, respectively). He returned to Canada as an assistant professor at the University of Alberta, which became his long-term base for both teaching and Arctic research. This simple statement disguises an enormous lifetime contribution of forty-eight years of groundbreaking fieldwork across the length and breadth of the Canadian Arctic.

Following completion of his postgraduate degrees, based on fieldwork in the vicinity of Broughton Island, John transferred his attention from Baffin to Ellesmere Island. From this point, his contributions can be divided into three phases: Ellesmere and the High Arctic; the northwestern section of the hitherto presumed outer limits of the Laurentide Ice Sheet; and the vast region of the also presumed non-glacierized region of the Beaufort Sea, in northern Alaska, and the farthest reaches of Canada's northwestern islands. Much of this work was undertaken in conjunction with several eminent Arctic colleagues and more than forty students, many of whom have themselves become leading contributors in unravelling the complexities of the geography and history of our northern landscape. Specific results of the highest importance include the mapping of, history of fluctuations of, and sea level relationships of the Innutian Ice Sheet—an unravelling of the 10,000 to 5,500–year record of penetration of driftwood into far northern fiords and bays from Siberia and its sudden blockage over five thousand years ago. This established the age of formation of the Ellesmere northern shelf ice, also vital to current evaluation of the impacts of climate change in the Arctic. A third contribution is John's total redrawing of Laurentide Ice Sheet extent in the far northwest and recognition of the catastrophic collapse of a former ice shelf that was over 100,000 square kilometres in extent, by far the largest ever known and an invaluable proxy for present-day ice sheet collapse in the Antarctic.

John's work has been recognized in numerous quarters. He was awarded an NSERC Northern Research Chair, renewed to cover the period from 2002 to 2012. He has been invited to lecture in the United Kingdom, Norway, Sweden, Iceland, Belgium, the United States, and across Canada. He was the main protagonist in the creation of Canada's northernmost national park—Quttinirpaaq, northern Ellesmere Island—and has devoted much time to both classroom and field teaching of Inuit and Gwitch'in students from Inuvik and Iqaluit. John continues a long commitment to lobbying for the establishment of an effective Canadian polar policy. He was inducted into the Royal Society of Canada in 2012. Currently he is a professor emeritus in Earth and Atmospheric Sciences at the University of Alberta.

BARRY GOODISON

Barry Goodison joined the Baffin Island project in May 1965 as a first-year undergrad from the University of Waterloo. His first task was to measure snow accumulation over the northwest part of the Barnes Ice Cap and then to work with Mike Church on fluvial geomorphology at Lewis River. Little did he realize at the time that this early encounter with the North, snow, and ice would set his career path in hydrometeorology and climate and cryosphere studies for almost fifty years. He joined Mike for two more summers in Baffin before conducting studies on Peyto Glacier in Alberta.

Barry completed his MA and PhD in geography at the University of Toronto under the guidance of

Dr. Ken Hare and Dr. Vit Klemes respectively. He joined Environment Canada as a research scientist in 1973, which led to an exciting career in research and management of national and international cold region science and technology for over forty years. His research included the development of new knowledge and technology for the measurement of snowfall and snow cover using advanced satellite and ground-based methods. For World Meteorological Organization (WMO)'s first major international experiment determining the accuracy of solid-precipitation measurements throughout the world, Barry was lead scientist. He also conducted and managed national and international field research studies for the development and validation of satellite algorithms and land surface process and climate models, including Boreal Ecosystem-Atmospheric Study (BOREAS) and Boreal Ecosystem Research and Monitoring Sites (BERMS) in Western Canada. For fifteen years he served as lead scientist for the Cryosphere System in Canada (CRYSYS) project, a Canadian contribution to NASA's Earth Observing System (EOS) program. The project brought together the cryospheric community distributed across Canada and, through the efforts of the participants, provided a training opportunity similar to that provided by the Baffin Island project.

Barry was active in the development and management of several national and international polar observation and research activities, many related to the International Polar Year (IPY) 2007–2008. He was the IPY coordinator for Environment Canada, was a member of the Canadian National IPY Steering Committee, and served on the World Meteorological Organization's Intercommission Task Force for IPY. Over the last five years, he led development of the concept and implementation strategy for the IPY legacy initiative, Global Cryosphere Watch (GCW), which is a new WMO involvement to provide authoritative, clear, and useable data, information, and analyses on the past, current, and future state of the cryosphere. With the Arctic Council's Arctic

Monitoring and Assessment Programme (AMAP), he has been involved in the development and completion of several collaborative polar research and planning initiatives, including, most recently, the Snow, Water, Ice, and Permafrost in the Arctic (SWIPA) assessment. Barry chaired the World Climate Research Programme's Climate and Cryosphere Project (WCRP/CliC) from 2002 to 2008 and is currently a member of the Global Climate Observing System (GCOS) steering committee.

Barry has authored or co-authored over one hundred papers and contributed to or edited numerous books, chapters, and agency reports, made countless presentations, and been a member of or chaired numerous Canadian and international committees and working groups. He continues to serve as an expert in the review of research proposals and projects. He has received national and international awards for his contributions. He was one of fifty recipients of the University of Waterloo's Fiftieth Anniversary Alumni Award (2007) and was recently awarded the Patterson Distinguished Service Medal (2012) for outstanding service to meteorology in Canada.

These achievements would not have occurred if it were not for the opportunity to be part of the Baffin Island project and to have the chance to work with and learn from the team of outstanding scholars who contributed to this early northern adventure.

WILLIAM (BILL) RANNIE

When I was selected for the Baffin Island project in 1963, as a first-year geography student at Queen's University, I was pleased, excited, and a little apprehensive. I expected it would be an adventure, but I had no idea then how my life's path would be altered and how lucky I was to have been chosen. The civil service application required that field assistants be able to "carry heavy packs over rough terrain" but instead of being a beast of burden, I had the much better fortune to be Mike Church's assistant, measuring

the water and sediment discharge of the Lewis River on the northwest corner of the Barnes Ice Cap. The two summers I spent with Mike, learning the basics of field hydrology (and a great many other things besides), were the best educational experience I could have had. Most importantly, we learned to use the salt dilution technique introduced by Gunnar Østrem to measure extremely turbulent discharge (for the first time in North America). This was a process made all the more challenging by flow conditions that were decidedly more difficult than had been anticipated. After two summers on the Lewis River, luck smiled again when I was given responsibility for stream gauging as part of the Decade Glacier project, for two more summers in beautiful Inugsuin Fiord.

The immediate result of this research in the North was a complete switch in my undergraduate focus from a vague notion about urban planning to physical geography, with a particular interest in hydrology and sediment transport. These interests carried through two graduate degrees and a forty-year career at the University of Winnipeg. While circumstances and opportunities changed the specifics of my research interests, a focus on hydrology remained.

All of this was the direct result of the opportunity provided by the Geographical Branch's Baffin Island project. Whatever path I might otherwise have taken, it would have been completely different without the Baffin experience and it's hard to imagine that it would have been more fulfilling. In the four summers on Baffin, I was privileged to meet so many superb scientists, to experience an astonishingly beautiful part of the country that would otherwise have been completely inaccessible, to receive an outstanding education almost without realizing it, and to have countless adventures all the while. In retrospect, I have often thought that the most amazing aspect of the project was the extraordinary amount of responsibility the Geographical Branch was willing to invest in such a junior person—and not just me, of course, but many others as well. This truly was a golden time to be a student, one that will probably never be repeated. Field research projects were being undertaken by government agencies and universities all over northern Canada. These projects provided opportunities for countless undergraduates, giving them skills and experience, changing many of their career paths, and generating an unintended but real added value beyond the scientific outcome. Such opportunities are rare now, as government withdraws from field research and pure science year by year and funds from other sources diminish. I was truly fortunate to have been part of the Baffin project.

JUNE (RYDER) CHURCH

June Ryder worked on Baffin Island for the Geographical Branch in the summers of 1966 and 1967, investigating alluvial fans and talus slopes. Interest in alluvial fans carried over into her PhD thesis—on paraglacial fans in the valleys of south-central British Columbia—completed at UBC in 1970. Her article in the *Geological Society of America Bulletin* (1972) on "paraglaciation," influenced by her Baffin Island experience, has spawned an entire subfield of geomorphological study. She subsequently worked for the Geological Survey of Canada for several summers under the direction of Dr. R. J. Fulton, carrying out terrain mapping, while teaching part time in the geography department at UBC during the remainder of the year. This in turn led to an appointment with the B.C. Ministry of Environment as head of the surficial geology unit from 1976 to 1981. Here she made major contributions to the development of the original British Columbia terrain classification scheme (1976); later she was responsible for revising and upgrading this system. Between 1982 and 2003 June worked as a geological (terrain) consultant specializing in terrain mapping and analysis for forestry (primarily slope stability). Provincial responsibilities included positions as member of the Clayoquot Scientific Panel (responsible for slope stability) and as provincial reviewer, responsible for quality control of terrain stability mapping (funded by Forest

Renewal B.C.); June also prepared the latest edition of the "Terrain Classification System for B.C." and compiled *A User's Guide to Terrain Stability Mapping*. She also conducted numerous workshops on this topic at locations around the province for government and private logging companies. In addition, between 1990 and 2003 she employed more than twenty of her former students, providing on-the-job training in air photo interpretation, terrain mapping and interpretations, slope stability assessments, and bio-terrain mapping. Though retired since 2003, June now works almost full time: as a volunteer bird surveyor for Bird Studies Canada and as editor of the British Columbia Field Ornithologists newsmagazine.

PATRICK JOHN (PAT) WEBBER

Being born in England in 1938, I was a pre-war baby. My formative years were spent roaming the Chiltern Hills in winter and spring and Exmoor every summer. My father was a Bedfordshire schoolmaster who knew every wildflower, and my grandfather farmed on the Devon side of Exmoor. My family never worried that I was gone from morning to evening with my dog—and my haversack containing oat cakes and collecting gear for birds' eggs and Lepidoptera. Later, I learned sociability and teamwork on the Baffin Island expeditions of the 1960s; in fact, teamwork became my lifelong mantra.

Following my bachelor's degrees in honours chemistry and botany from the University of Reading, I applied to Queen's University in Kingston to pursue my interest in the natural history of swamps and bogs. To my surprise, Roland Beschel offered me a teaching and research scholarship. I arrived in Canada in the summer of 1960 to find that Beschel was on Axel Heiberg Island; he had left me instructions to learn the local flora and to identify a topic for a master's dissertation. Beschel soon discovered that I had been bitten by the polar bug through the romance of Robert Falcon Scott's trek to the South Pole and my undergraduate professor's (T. M. Harris) work on the Jurassic flora of Scoresbysund (Ittoqqortoormiit), East Greenland. Beschel made it clear that if I could demonstrate research skill working in southern Ontario bogs and swamps he would help me find a way farther north. Through the foregoing serendipity and the Geographical Branch's discoveries of fossiliferous deposits and rock lichen patterns, Beschel made good on his promise when he negotiated with Jack Ives a place for me on the Baffin expeditions.

So, in May 1963 I found myself helping Gunnar Østrem dig ice from a moraine, and for the rest of the summer I learned the basics of periglacial geomorphology from John Andrews and Mike Church in exchange for tips on vascular plant identification and lichenometry. My everlasting gratitude is to Jack Ives and his team for the world-class field support and complete freedom to pursue my own science.

My doctoral dissertation theme was exploration of the nature and structure of High Arctic plant communities. At that time there was much debate, even rancour, about the level of organization of species in vegetation in general; the common wisdom was that, in the Arctic, competition among species was so minimal that repetitive assemblages suitable for mapping or replicable experimentation did not exist. So I set out to test and compare for R. H. Whittaker's alternate hypotheses: the individualistic hypothesis and the association unit hypothesis. I pioneered the application of ordination and dendrogram classification methods to Arctic vegetation. Most of the 1963 field season was spent collecting plants and establishing—with John Andrews and every member of the expedition—a network of lichen-measuring stations throughout the Flitaway Lake and Lewis Valley areas. In 1964 I sampled eighty-nine plots to serve as the basis for testing the Whittaker hypotheses. Over the years they have been resampled several times. During the Fourth International Polar Year, these plots and several of the Baffin Island lichen stations were resampled under the banner of "Back to the Future."

In 1966 I joined the faculty at York University and developed a field program on the southern shore of Hudson Bay, and in 1967 I rejoined the Geographical Branch team for a summer at Kangirlugag Fiord, which is adjacent to Ekalugad Fiord. In 1969 I joined the University of Colorado's INSTAAR and participated in many large ecological projects, such as the International Biological Program, Tundra Biome, the San Juan Ecology Project, and the Alpine Long-Term Ecology Project. Over the years I spent considerable field time on the North Slope of Alaska and, in particular, developed mapping methods that have been used to monitor long-term cumulative changes in Alaska's largest oil field, at Prudhoe Bay. Throughout my career I mentored a fair number of graduate students and thoroughly enjoyed teaching large general ecology and introductory botany classes. I was director of INSTAAR from 1980 to 1986. I served on a number of national and international committees, for example, the U.S. National Polar Research Board, Man and the Biosphere Program, and the International Arctic Science Committee, where I was president from 2001 to 2006. I also served in Washington, D.C., at the U.S. National Science Foundation as director of the Ecology Program (1986–1989), director of the Arctic Systems Science Program, and head of the Arctic Section of Polar Programs (1993–1995). In 1990 I joined Michigan State University where, for a while, I directed an experimental dairy farm and the W. K. Kellogg Biological Station; I was later able to return to my own Arctic research, work with another generation of graduate students, and in particular, establish the Alaskan group of the International Tundra experiment. In 2005 I retired to take up farming and woodworking in New Mexico. I was in 2010 the first recipient of the International Arctic Science medal, awarded in Oslo, for lifetime contribution, leadership, and mentoring in Arctic science.

Geographical Branch Staff

JOHN T. ANDREWS

John was born in 1937 in Millom, England. He completed his bachelor's degree in geography at the University of Nottingham in 1960. Following graduation, he emigrated to Canada, spending his first Canadian winter at the McGill Sub-Arctic Research Laboratory. There he prepared for his master's degree fieldwork on the Labrador coast north of Nain while at the same time beginning a study of lake ice growth rates in central Labrador-Ungava. John completed his MA in geography at McGill in 1961, after which he served as a graduate assistant in north-central Baffin Island for the federal Geographical Branch. Becoming a permanent member of the branch staff in 1962, he concentrated his efforts from 1962 until 1967 on glacial geomorphological research and sea level fluctuation. During this period, he took a year's sabbatical leave that enabled him to complete his doctoral degree at the University of Nottingham, in 1965, under the supervision of Dr. Cuchlaine King.

In 1968, John was appointed associate professor in geological sciences and assistant director of the Institute of Arctic and Alpine Research (INSTAAR), University of Colorado, Boulder. He devoted the remainder of his long career to Arctic research from his base in Boulder. In 1975, John received the Kirk Bryan Award of the Geological Society of America for his monograph *A Geomorphological Study of Post-Glacial Uplift with Particular Reference to Arctic Canada* and, in 1978. was awarded a DSc degree by the University of Nottingham. The University of Colorado recognized him as Distinguished Research Professor.

John advised and supported numerous graduate students and published several hundred papers in a wide range of scientific journals. His mature research is best described as a study of terrestrial and marine glacial systems. This work recognizes the progressive expansion of his fieldwork from the eastern Canadian Arctic to the fiords and continental shelves of Greenland, Baffin Island, Labrador, and Iceland. Since 1970, John has been the dominant figure in the highly effective annual series of Arctic Workshops that originated at INSTAAR in Boulder and has extended to alternate-year locales in Canada, the eastern United States, Norway, and Iceland. He has developed longstanding research collaboration with the Atlantic GeoScience Centre (Canada), the universities of Tromso and Iceland, and GeoMar (Germany). In 2002, the American Geological Society recognized John's extensive contribution by arranging for a special session in his honour during its annual meeting in Denver. He is probably the single most prolific scientist in his field.

LYN (DRAPIER) ARSENAULT

When I was in my final undergraduate year at Southampton University (hons. botany and geography, 1966), Dr. John Andrews visited Dr. Roger Barry (my academic supervisor) and gave an inspiring lecture on the Arctic field research ongoing at the Geographical Branch in Canada. As a fervent devotee of hands-on mapping, fieldwork, and camping, I promptly applied to Ottawa! With job prospects bleak in the U.K. and it being almost impossible for a woman to be considered in several other countries, it was a relief to be offered a position as research assistant, starting in September 1966. Roger Barry joined the Geographical Branch the following year for his sabbatical leave. I worked from 1966 to 1968 in the branch's Division of Physical Geography and spent a memorable two months on Henry Kater Peninsula, Baffin Island, in summer 1967 with Dr. Cuchlaine King and Mary Strome. Following the breakup of the branch, I consulted for the Geological Survey from 1968 to 1970, and from 1971 to 1973 I worked for the Glaciology Division of Environment Canada.

My career to this date had involved air photo interpretation of northern land features, but then came the challenge facing Canada as economic interests necessitated locating and identifying sea ice and icebergs year round. Thus, I founded a consulting business, Cold Regions Remote Sensing, in 1973 and for thirty years provided information on sea ice and iceberg regimes, and on a wide range of remote sensing imagery for their detection and discrimination. This included research for aircraft and shipping companies, the petroleum industry, and government departments (Environment, Fisheries and Oceans, Natural Resources, Transport). I also worked as a scientific editor.

I began analyzing sea ice extent with some of the earliest optical Landsat data in 1973, in cloud-free daylight conditions, and ended with superb imagery from Canada's Radarsat, the first civilian, spaceborne synthetic aperture radar, which can distinguish sea ice types and detect icebergs under all weather conditions, both day and night. The imagery analysis involved extensive fieldwork in the Canadian Arctic and southern Greenland: on the sea ice surface, aboard ships, and on Canadian, American, and Danish research aircraft.

I served on several committees and boards over the years: Canadian Advisory Committee on Remote Sensing/Working Group on Ice (secretary); Canadian Ice Working Group (co-chair); Alliance for Marine Remote Sensing; Ottawa Glaciological Group (vice-chair) of the International Glaciological Society; and The Arctic Circle (president).

Since retiring, my continued enthusiasm for the Arctic has prompted extensive camping trips throughout the North with my husband, Jim.

ROGER G. BARRY

Professor (geography) since 1971; distinguished professor since 2004, University of Colorado. Director since 1976, National Snow and Ice Data Center/World Data Center for Glaciology, now employing one hundred staff, archiving and distributing snow and ice data products (http://nsidc.org).

Teaching/researching climate change, Arctic and mountain climates, snow and ice processes; field research in the Canadian Arctic, New Guinea. Published over twenty textbooks, over two hundred articles; supervised fifty-five graduate degrees. Awards totalling $78 million as Principal Investigator (PI) or co-PI since 1996. Service to national committees (NAS), international programs (WCRP, GTOS, IPCC).

Guggenheim Fellow (1982–1983); Fulbright Teaching Fellow, Moscow (2001); Fellow, American Geophysical Union (1999); Foreign Member, Russian Academy of Natural Sciences (2001). Awarded the Goldthwait Polar Medal (2006), and the Royal Geographical Society Founder's Medal (2008). Visiting professor in Australia, France, Germany, Japan, New Zealand, Russia, Switzerland, and United Kingdom.

Since high school, I have had a keen interest in other cultures and languages. Born in England, I went on three student exchanges to France as a teenager. My master's degree in climatology from McGill University included a year at a the McGill Sub-Arctic Research Laboratory. After seven years at the University of Southampton, including two seasons of meteorological fieldwork in Arctic Canada, I moved to the University of Colorado in 1968. I have trained over fifty graduates in climatology, snow and ice, and mountain and polar environments.

I became director of the World Data Center for Glaciology in 1976 and the National Snow and Ice Data Center in 1982. They grew from a two-person grant to a scientific and technical staff of one hundred with funding of $8.5 million/year; the combined activities span snow cover, floating ice, glaciers, ice sheets, and frozen ground/permafrost. The center has responsibility for snow and ice remote sensing and *in situ* data products under NASA's Earth Observing System and it archives polar data for NSF and NOAA.

Sabbatical leaves/visits have allowed me to lecture at institutes in Europe and Russia (I am fluent in French, German, and Russian) and in China, Japan, and Venezuela. I am a member of international research and data committees with the World Climate Research Programme, Global Digital Sea Ice Data Bank, Global Terrestrial Observing System, and International Permafrost Association and review editor for the Intergovernmental Panel on Climate Change.

Graduate education: I trained some of the leading climatologists of the present generation: Ray Bradley (former chair of geosciences, UMass), Jill (Jaeger) Williams (former director, IHDP), Wally Brinkmann (former chair of geography, University of Wisconsin–Madison), Gerry Meehl (NCAR scientist), Ells LeDrew (former dean of environmental sciences, University of Waterloo), Henry Diaz (NOAA-CDC scientist), George Kiladis (NOAA-PSD scientist), Roger Pulwarty (scientist, NOAA, Boulder), Mark Serreze (NSIDC scientist), Jeff Key (scientist, NOAA-CIMMS, Madison), Andrew Carleton (Penn State University), Jeff Rogers (Ohio State University), Mark Anderson (University of Nebraska), Shari Fox-Gearheard (CIRES visiting fellow, Clyde River, Nunavut).

Supervision of Jill Williams's PhD entailed the first use of the NCAR general circulation model to simulate global climate with Glacial Maximum ice margins and changed sea surface temperatures. Mark Serreze and Jeff Key worked on Arctic climate and sea ice. Jeff Rogers (with Harry van Loon) documented the North Atlantic Oscillation.

Research on the climate of Arctic and Subarctic Canada; mountain climates in Colorado, New Guinea, and Venezuela; paleoclimate of the Last Glacial Maximum and ice sheet inception, changes in snow cover, sea ice, mountain glaciers, and frozen ground.

Group leadership of projects on landfast sea ice in Baffin Bay and the Beaufort-Chukchi seas, Arctic climate–sea ice interactions, passive microwave remote sensing of Arctic sea ice. Data rescue programs with the Institute of Geography, RAS, Moscow; the Institute of Fundamental Biology and Soil Science, RAS, Pushchino; the Arctic and Antarctic Research Institute, Saint Petersburg; and the WDC for Glaciology, Lanzhou.

Published books include: eight editions and four translated editions of *Atmosphere, Weather and Climate* (with the late R. J. Chorley, 1968–2003); two editions and two translated editions of *Mountain Weather and Climate*; *Synoptic Climatology* (with A. H. Perry, 1973); *Synoptic and Dynamic Climatology* (with A. M. Carleton, 2001); *Arctic and Alpine Environments* (co-edited with J. D. Ives, 1974); *The Arctic Climate System* (with M. C. Serreze, 2005); *The Global Cryosphere: Past, Present and Future* (with Thian Yew Gan, 2011); *Essentials of the Earth's Climate System* (with E. A. Hall-McKim, 2014).

JANE (PHILPOT) BUCKLEY

Raised by Oxbridge scientists in Oxford, I studied at Cambridge, graduating in 1962 with a degree in geography. I was scientific leader for the first British Girls Exploring Society expedition (Lofoten Islands, Norway) before coming to Canada. After a year teaching geography in Ottawa, I was fortunate to be recruited by Dr. Jack Ives as an air photo interpreter in the Geographical Branch. In 1964, I shortened a world trip to return to Ottawa, tempted by the possibility of going to the Arctic.

So in 1965, after rifle practice with the RCMP (who suggested that I take up shooting), I served as field assistant to Dr. Cuchlaine King working on raised beaches and glacial features on Baffin Island. Practical plane-tabling sessions in Cambridge had failed to teach that squishing mosquitoes messes up maps and that hands that hold survey poles should not be covered in repellent—the numbers on the poles were soon illegible!

In 1966, assisted by Penny Crompton, I worked on the coast near Sam Ford Fiord. After our radio failed, I learned always to check my own equipment. Despite going on half rations until we were rescued by the "Ives Air Force," I put on weight that summer. On my return to Ottawa to marry Michael Buckley, my wedding dress had to be altered—I blamed rum-soaked fruitcake!

After several co-authored publications based on the Baffin fieldwork, I completed a paper on glacier gradients, comparing those in Baffin with those in Greenland and Antarctica. This work ceased in 1967 and I was transferred to the Geological Survey.

After the birth of a daughter, I learned scientific editing and became sole editor for the Forest Management Institute for four years. In the mid-1970s, I helped establish the Freelance Editors' Association of Canada (now the Editors' Association of Canada). Government departments provided work on things environmental, icy, oily, and fishy. I learned how powerful the legal pen is when applied to documents: my job was to stitch up the remains and make them read like science. Editing books and conference proceedings gave me great pleasure. Annual conferences of the Council for Biology Editors took me across North America.

In 1988 I joined Agriculture and AgriFood Canada, editing award-winning research books for its Research Program Service (RPS). When offered early retirement in 1995, I revived Gilpen Editing Service, which I operated until 2003. While fascinating topics had ranged from flies to fescues and from statistics to soils, the RPS team had not the same rapport as I had experienced at the Geographical Branch. Thanks for the memories, Jack.

Currently, I copyedit the local community newspaper; prepare the annual Conservation List for, and offer sage advice to, Rare Breeds Canada (RBC); learn about handling agility dogs; count birds for the local Project FeederWatch; and garden passionately whenever my garden is not under snow. I will be a member of the Arctic Circle committee for the next three years, where cherished memories of the Baffin Island adventure are rekindled.

GEORGE FALCONER

Born in U.K., 1929. Served in British Army of the Rhine (1947–1949), then read geography at Cambridge University. Participated in the geography department's glaciological work in the Jotunheim Mountains, Norway, and later, privately, in the Swiss Alps. Emigrated to Canada in 1953. Worked on terrain analysis and air photo interpretation keys in Coppermine, N.W.T., coastal areas for Geographical Branch (1954). Special assistant to Professor J. Tuzo Wilson, Geophysics Laboratory, University of Toronto (1955–1958). Managed and participated in extensive air photo terrain analysis of much of Arctic and Northern Canada, culminating in the publication of the first Glacial Map of Canada (co-author, designer, and production manager). Organized and participated in University of Toronto geophysical and glaciological expeditions to the Salmon Glacier, B.C. (1956–1957).

After a brief experience in town planning in Toronto, joined the Geographical Branch in 1958. Worked on area economic surveys on the east coast of Hudson Bay (1959–1961). Participated in the Baffin Island research project (1962–1965). In a change of focus, transferred to the Geographical Branch's National Atlas program in 1967. Initially in charge of the physical geography component of the atlas. Later served as special assistant to director-general of Surveys and Mapping Branch and secretary of Interdepartmental Committee on Remote Sensing.

Appointed chief of Geography Division, Surveys and Mapping Branch, and editor-in-chief, National Atlas of Canada (1974). Directed and edited the final publication of the fourth and fifth editions of the national atlas and numerous other publications including the *Canada Gazetteer Atlas*. Served as secretary of the National Advisory Committee for the National Atlas and as senior advisor, National Geographical Information. In 1990 left government service to pursue interests in visual arts and art history.

CUCHLAINE A. M. KING

Cuchlaine was born in 1922 and grew up amidst academics in Cambridge, England (her father, Professor W. B. R. King, FRS, was professor of geology). She obtained her BA in geography in 1942 and entered the Women's Royal Naval Service. After the war she returned to Cambridge for her doctoral degree, awarded in 1949. In this she specialized in shoreline and beach processes, an element of physical geography that she pursued throughout her career. Her 1959 book *Beaches and Coasts* won the admiration of Professor William Balchin, who described her work as a "unique approach with a mathematical and quantitative treatment"; this was at an early stage in what became known as the "quantitative revolution."

Apart from a year at Durham University, her university career centred on the University of Nottingham, which—with her undergraduate students Jack Ives, John Andrews, and Roger Tomlinson—came to have a remarkable impact on Canadian Arctic research. She accompanied Jack Ives to Iceland in 1953 and 1954, thereby initiating a long-term involvement in glaciology, after which she worked for several years with Vaughn Lewis and Cambridge colleagues on an intensive study of the Austerdalsbreen in the Norwegian Jotunheimen. She joined the Baffin Island project for two summers (1965 and 1967) and provided the psychological assist that was vital in pressuring the Canadian government to accept the notion of gender equality in fieldwork in the Canadian Arctic. She was promoted to reader in geography in 1962 and full professor in 1969. She joined Jack and Pauline Ives for the 1968 INSTAAR summer research and teaching program in the Colorado Front Range. Thereafter, she travelled and published extensively and is credited with a library shelf of major textbooks, including *Glacial and Periglacial Geomorphology*, which she authored jointly with Chris Embleton.

Cuchlaine was one of very few women to hold a lifelong academic career beginning shortly after the Second World War. She and Dr. Moira Dunbar

(Canadian Defence Research Board) produced all but one of the articles by women published in the *Journal of Glaciology* during its first two decades.

OLAV LØKEN

Olav was born on February 23, 1931, in the Norwegian town of Alesund. After completing his early degrees at the University of Oslo he worked as a glaciologist at the Wilkes Station in Antarctica as a member of the U.S. Antarctic Program during the International Geophysical Year (1957–1958). From there he spent a year at the McGill Sub-Arctic Research Lab. This was followed by fieldwork in the Torngat Mountains of northern Labrador that formed the basis of his McGill doctorate. His first postdoc appointment was as assistant professor in the department of geography at Queen's University.

1964: Joined the Geographical Branch as chief of the Division of Physical Geography and, in addition to directing the development of the division, devoted a large part of his energies to the Baffin Island expeditions, 1964 to 1967.

1967 to 1976: Became head of the Glaciology Subdivision in the Inland Waters Branch, Department of Energy, Mines, and Resources. Much of his work focused on projects launched in the context of the International Hydrological Decade (1965–1974). He also continued his earlier work along the Canadian eastern seaboard, particularly on submarine geomorphology and the reconnaissance survey of the Baffin Island fiords.

At the end of this period he was seconded to the Environmental and Social Program, Indian and Northern Affairs, to investigate the environmental and social impacts of the Polar Gas pipeline.

1976 to 1984: Became director of the Northern Environmental Protection Branch, Indian and Northern Affairs, where he directed, planned, and coordinated environmental assessments of major development projects according to the federal Environmental Assessment and Review Process. This included environmental assessments of the Norman Wells Oil Field Expansion and Pipeline Project, Mackenzie Delta gas projects, the Arctic Pilot Project, and the Beaufort Sea Hydrocarbon Production Proposal.

Oil and gas exploration in the Eastern Arctic was about to begin. The government, in cooperation with industry, established the Eastern Arctic Marine Environmental Studies (EAMES) program to collect and interpret environmental data in order to improve understanding of the ecosystem in anticipation of expanding oil and gas activities. Studies were conducted in Baffin Bay and Lancaster Sound.

The branch was also involved in the Arctic Land Use Research Program, investigating the rehabilitation of sumps at earlier drilling sites and the effect of vehicle traffic on the tundra surface.

1981: Headed the Task Force on Beaufort Sea Developments. He planned and directed the work of an interdepartmental and intergovernmental task force established to examine the state of government preparedness in the Beaufort/Delta area. The task force report recommended policy and program initiatives, and a formal government program was established.

1984 to 1989: Was director of the Environmental Studies Research Funds (ESRF), Indian and Northern Affairs/Energy, Mines, and Resources, where he was accountable to the Environmental Studies Management Board for the scientific, administrative, and financial planning of the ESRF's operation. This targeted research program facilitates decision-making regarding oil and gas activities in Canada's frontier regions. Funded by industry, the program involved extensive contacts with private sector companies and First Nation and Inuit groups.

1989 to 1990: Served as scientific advisor to the Polar Continental Shelf Project, Department of Energy, Mines, and Resources.

1991 to 2005: After retiring from the Government of Canada, Olav worked with the newly created Canadian Polar Commission to promote Canadian

involvement in Antarctic-related research. Among his many activities, he was secretary of the Canadian Committee for Antarctic Research from its inception in 1998 to 2005. He was instrumental in Canada's application to join the Scientific Committee for Antarctic Research (SCAR) and represented Canada there. He was also active in efforts leading to the passage of the Antarctic Environmental Protection Act in 2003.

Miscellaneous: In 2001 Olav served as an observer for the International Association of Antarctic Tour Operators (IAATO) on an Antarctic cruise ship. He has been a member of a number of job-related and professional committees at the national and international level. He served as president of the Canadian Nordic Society and as president of the Arctic Circle, both in Ottawa.

GUNNAR ØSTREM

Gunnar was born in Oslo in 1922. He obtained his Cand. mag. degree at the University of Oslo in mathematics, chemistry, and geography; this was followed by the Cand. real. degree in 1954 (geography with specialization in glaciology). He was awarded the Filosofie Licentiat in 1961 and the Filosofie Doctor in 1965, both at the University of Stockholm with specialization in glaciology. His early career involved teaching at high schools and university in Norway and Sweden, and he served as assistant professor of physical geography at the University of Stockholm from 1958 to 1962.

Gunnar joined the Geographical Branch and began research on the Baffin Island project in 1962, serving as head of the Glaciology Section until 1966 and returning to Stockholm later in 1966 as associate professor and then, from 1981 to 1983, as full professor. He was head of the Glaciology Section at the Norwegian Water Resources and Energy Board (NVE), Oslo, from 1962 to 1981, with interruptions in Stockholm and Ottawa and a sabbatical at Carleton University (1971–1972). After 1981 he continued part time with NVE until his retirement in 1992.

The Hans Egede Medal (the highest award of the Danish Geographical Society) was awarded to Gunnar in 1982 for "extensive research in Arctic areas"; he was also honoured with the King of Norway Gold Medal in 1992 (for scientific achievements in Norway and abroad) and the Swedish J. A. Wahlenbergs Silver Medal in 1993 (for seventeen years of editorship of the scientific journal *Geografiska Annaler*).

Gunnar made extensive contributions in training and research, and developed and taught specialized courses in glaciology, for various United Nations agencies (e.g., UNESCO, WMO, FAO) and especially in the Indian Himalaya; he was invited to give summer courses and field training in Canada, Nepal, Pakistan, Austria, Argentina, and Greenland. He published more than eighty scientific and technical articles and books and many specialized maps related to water resources, hydropower, and glacier meltwater and runoff.

BRIAN SAGAR

Brian was born in 1927 in Colne, Lancashire, England, and earned a BSc in geography at Hull University. After spending three years in Africa in an administrative capacity with the United Africa Company, and following a brief return to England, he emigrated to Canada in 1958 to join the Defence Research Board IGY summer expedition to northern Ellesmere Island. So began his scientific passion for the world of snow and ice. His glacier mass balance studies, central to his McGill University MSc (1959), involved further Arctic research with the Defence Research Board and AINA, and he eventually joined the Geographical Branch in December 1961. This led to his "command" of the Barnes Ice Cap as a key member of the branch Baffin Island expeditions, from which he took several short leaves of absence to teach physical geography at Carleton University. His many friendships, cemented by the camaraderie of Arctic adventure and scientific endeavour, remained profoundly important for the remainder of his life. It was also in the Geographical Branch that he met Norma, leading to twenty-five years of marriage before his untimely death from cancer on January 12, 1990.

Brian left Ottawa in 1966 to take up a faculty position in the department of geography at Simon Fraser University (SFU), where he served as acting head of the department from 1969 to 1971. He taught physical geography, climatology, and the geography of natural hazards, together with a regional course on Africa, to thousands of undergraduates and graduates during his twenty-three years at SFU, together with field courses that included supervision of graduate students at both SFU and UBC. He also served as vice-chairman of the Vancouver branch of the Canadian Meteorological Society and as organizing secretary of the Canadian Meteorological and Oceanographic Society from 1976 to 1979. These interests were reinforced by numerous environmental and associated political activities. Brian was crucial to the evolution of the formal establishment of glaciology within the Canadian federal government.

[Author's note: Brian's brief biography was summarized from an obituary generously supplied by his surviving spouse, Norma.]

PORTRAITS
OF FIELD
TEAM MEMBERS

Fig. 66: Olav Løken

Fig. 67: Cuchlaine King

Fig. 68: David Harrison

Fig. 69: Rolf Feyling-Hanssen

Fig. 70: John Andrews

Fig. 71: John England

Fig. 72: Mike Church

Fig. 73: June Ryder

Fig. 74: Bill Rannie

Fig. 75: Doug Christian

Fig. 76: Jane (Philpot) Buckley

Fig. 77: Penny Crompton

Fig. 78: Jean Logie

Fig. 79: Lyn (Drapier) Arsenault and Mary (Strom) Millest

Fig. 80: Elizabeth from Clyde River

Fig. 81: Norman Gray

Fig. 82: Bruce Bradley

CHAPTER 11

ASSESSMENT OF
THE SCIENTIFIC RESULTS

Assessment of the scientific results of the 1961 to 1967 Baffin Island expeditions must be set within the context of the general knowledge and hypotheses that were prevalent during the 1950s and 1960s. Even in the 1960s, additional research was undertaken in neighbouring regions, such as Axel Heiberg and Devon islands, the Canadian Arctic mainland (Lee, 1960; Müller, 1962), and in southern Baffin Island (Blake, 1966). Similarly, the earlier research in Arctic Scandinavia (De Geer, 1912; Frodin, 1925; Gjessing, 1960; Hoppe, 1952, 1957; Mannerfelt, 1945; Schytt, 1949) and my own experience in Iceland (King & Ives, 1956) and with Olav Løken in Labrador-Ungava (Ives, 1960b, 2010; Løken, 1960, 1962) had greatly influenced the evolution of thinking within the Geographical Branch, as previously acknowledged. Much of this work also continued contemporaneously with the Baffin field program and, given the participation of our Scandinavian contingent, cross-fertilization was automatic. In the words of Pat Webber,

> The pursuit of knowledge, that is, science, usually progresses in small steps, each step building on what went before. It is reasonable to say that the Baffin Island expedition members and their students contributed a quantum leap to, and the foundation for, the present level of understanding of the ice age history, the periglacial environment, the Holocene climate, and the biota of the eastern Canadian Arctic. It is also fair to say that the group's early estimates of process rates and interpretations of glacial history have held up remarkably well. Nevertheless, it is extremely satisfying to see the great strides that have been made since the 1960s, especially in recent decades, with the development and application of new technology and methods. (Personal communication, February 20, 2014)

The first task of this chapter, therefore, is to outline what emerged from Geographical Branch efforts in Baffin Island within the stated context. A second task is to address, in retrospect, how the results of later research (1968 to present) have been influenced by the 1961–1967 studies and to what extent the latter have been modified and/or contravened. This is presented only in outline because of the very extensive publication record and the ongoing nature of the research effort. Thirdly, we need to assess the influence of the Baffin project on the careers of a substantial cohort of scholars in geography and related disciplines (see chapter 10).

Concept of Canadian Arctic and Subarctic glaciation prior to 1960

Early publications on the ice age history of the eastern Canadian Arctic and Subarctic were influenced by more detailed work in Fenno-Scandinavia. One of the more relevant hypotheses had been that, during the ice ages, large parts of the Norwegian coastal mountain summit areas had projected above the maximum height of the Fenno-Scandinavian Ice Sheet as *nunataks*; hence the "nunatak hypothesis."[1] The hypothesis was developed to account for the peculiarly restricted distribution of a large group of arctic-alpine plant species that many botanists argued must have survived the ice ages in (or close to) places where they flourish today. The nunatak hypothesis was strenuously contested by most earth scientists of the time, although this interdisciplinary confrontation was by no means absolute.

The Scandinavian dispute about the nunatak hypothesis persisted for almost a century and is still not entirely resolved (Brochmann, Gabrielsen, Nordal, Landvik, & Elven, 2003). The frequently astringent academic exchanges are divided among two distinct groups of researchers: biologists, principally botanists, with a few supporting zoologists, on the one hand, and earth scientists and most physical geographers on the other. The botanists had built an extremely detailed database on the peculiar distribution of the arctic-alpine group of plant species that seemed restricted to specific high mountain areas of Norway and Sweden with no clear means of inmigration following the disappearance of the major ice sheet (Dahl, 1955, 1961; Ives, 1974; Löve & Löve, 1963, 1974). The earth scientists appeared to have scant sympathy for their rivals, although the high mountain areas in question had revealed little, if any, unambiguous field evidence to support the argument that, at its maximum, the Fenno-Scandinavian Ice Sheet had completely overtopped all the high summits. Knowledge of the extent and significance of cold-based ice, however, was not well developed at the time. In contrast, the lower elevations contained widespread and unequivocal evidence of active glacial erosion and deposition (that is, by warm-based ice, as was understood much later). Frequently, this was marked by a distinctly aligned upper limit that sloped down along the main valleys and fiords toward the Norwegian Sea, often paralleled at lower levels by the lateral moraines of former glaciers. The proponents of the nunatak hypothesis took this upper limit as proof of the maximum extent of ice age glaciers (Dahl, 1955, 1961). Nevertheless, it was widely recognized that sections of the high "nunatak areas" had also supported small, thin ice caps and local glaciers. The search for support of the nunatak hypothesis was extended to Greenland, Svalbard, Iceland, and Alaska.

The controversy was also "exported" to northeastern North America, where the same lack of unequivocal mountaintop evidence for glacial erosion became a basis for continued controversy. However, in comparison to Norway and Sweden, the botanical evidence was sparse, simply because the eastern coastal mountains were virtually unexplored. Some botanists employed the nunatak hypothesis to explain what was known about arctic-alpine plant species distribution, principally in New England (e.g., Mount Washington, and Mount Katahdin: see Fernald, 1925). Contrasting sets of earth science data fuelled the argument, focussing on the Torngat Mountains of northern Labrador, northern Newfoundland, and the Shickshock Mountains of Gaspésie, Quebec. Daly (1902) and Coleman (1920, 1921, 1926) argued that the Torngat Mountain summits had remained above the maximum height of the Laurentide Ice Sheet (technically, the Labradorean Ice Sheet at that time). Their reasoning depended heavily on the absence of indications of glacial action on the summits and the assumption that development of the extensive high-level boulder fields (*felsenmeer*, or mountaintop detritus) would take an extremely long time. However, in 1933, Dr. Noel Odell (of Mount Everest fame) reported what he assumed to be definitive evidence of total submergence by ice of all the Torngat Mountains, based on his interpretation of faint glacial striations and his assumption that the mountaintop boulder fields were the product of rapid (i.e., postglacial) frost shattering after the disappearance of the final ice sheet. The time needed for formation of the extensive boulder fields (*felsenmeer*) became central to much subsequent dispute, although a lack of dating techniques at the time meant that much of the argumentation was a matter of personal opinion, if not downright specious.

A pre-1960 attempt to break the logjam developed from interpolation of the scattered knowledge of higher sea levels, based on the discovery far above present sea level of marine molluscs (seashells) and terraces similar to modern sea level beaches and deltas. It had long been understood, especially based on earlier work in Fenno-Scandinavia around the shores of the Gulf of Bothnia, that variations in the height of the marine limit was a reflection of regional variations in thickness of the former ice sheets. Here was a substantive argument based on the principal of isostasy—that the weight of the former ice sheets caused a proportionate depression in the earth's crust. After the ice sheet disappeared (indeed, even while it was thinning and retreating), the sea crossed the recently exposed land (which was rebounding to its pre–ice age levels), reaching elevations high above modern sea level and marking the uppermost shoreline (i.e., the marine limit), which often is sixty to eighty metres above modern sea level in the inner fiords of northeastern Baffin Island. But despite the fact that the sea level was rising as the ice sheets melted and retreated, the land eventually rebounded even more extensively, causing the relative position of sea level to lower (or regress) continuously back down to its modern position—this is so-called postglacial rebound.[2] Evidence for the highest former sea levels, at more than three hundred metres above present, is located in the southeastern sector of Hudson Bay; it is thought to mark the former maximum thickness of the Laurentide Ice Sheet and thus the location of

greatest unloading of the formerly (glacio-isostatically) depressed land. Prior to the Baffin Island surveys, however, the largely unexplored nature of the eastern Canadian Arctic provided only scattered observations, often limited to sightings from shipboard, of apparently horizontal terraces high above present sea level. When the technique of radiocarbon dating, especially applicable to seashells associated with the raised marine shorelines, became widely available in the 1960s, there was a great leap forward in understanding the interrelations between these raised marine shorelines and the regional history of retreat of the Laurentide Ice Sheet.[3]

Before the 1960s—the decade of the sea level surveys by the Geographical Branch across Baffin Island—the sparsely available evidence was absorbed by Professor Richard Foster Flint, who emerged as the doyen of North American glacial geologists during the period from 1940 to 1972 (Flint, 1943, 1945, 1947, 1957, 1971). He accepted the evidence presented by Odell (1933) from the Torngat Mountains as proof positive that the continental ice sheet had totally overtopped the highest summits. He also interpreted the highly equivocal observations of Wordie (1938) in the fiords of northeastern Baffin Island as ostensibly recording very high former sea levels; this, of course, facilitated Flint's presumption of extremely thick glacial ice sufficient to overtop the highest coastal summits.[4] Thus, he developed a model for the growth and decay of the Laurentide Ice Sheet that became the prevailing paradigm, receiving almost universal endorsement.

Flint referred to his concept of North American glacial history as a "mirror-image model" of the pre-existing Scandinavian theory of growth, climax, and decay of the Fenno-Scandinavian Ice Sheet. This mirror-image depiction, derived from the presumed similarity of the topography on either side of the North Atlantic Ocean, augmented by the much more advanced status of relevant Scandinavian research, could also be directly imported. Flint's vision of east coast Canadian topography was vital to his hypothesis. But while the outlines of the coast had been reasonably well mapped, vast areas of the interior were practically unknown. Nevertheless, he assumed that the eastern coastal areas, from northern Ellesmere Island down the extent of Baffin Island to mid-Labrador, supported high mountains facing Greenland, Baffin Bay, Davis Strait, and the western Atlantic, while their western flanks sloped down steeply to a series of inland plateaus and lowlands leading to Foxe Basin and Hudson Bay.[5] This, in effect, was the mirror image of the high coastal mountains of Norway that sloped eastward across Sweden to the Gulf of Bothnia.

Flint presumed, moreover, that the Gulf of Mexico was the origin of atmospheric low pressure systems that moved northward and northeastward, providing the moisture source for the growth of the Laurentide Ice Sheet once air temperatures began to fall with the onset of an ice age. As Flint's hypothetical snowline lowered with the falling air temperature, it would eventually first intersect the summits of the coastal mountains and so cause the buildup of glaciers and small ice caps. The masses of ice would thicken and spread, forming glaciers that flowed eastward down the fiords and out into the Atlantic Ocean and Baffin Bay, breaking off as icebergs. This would place a check on the increase in accumulation at the higher elevations. In contrast, those glaciers flowing down the "western flank" of the mountains would build up into vast piedmont lobes on the plateau surfaces, a process further augmented by the fact that they were expanding into the source of solid precipitation—the moist air masses flowing from the Gulf of Mexico. This was Flint's so-called model of "highland origin and windward growth." Eventually, the accumulation of ice to the west of the coastal mountains would exceed them in elevation, causing a reversal of flow through the mountains and into the Atlantic, Baffin Bay, and Davis Strait, submerging the highest summits in the process.

Continued westward growth of the ice sheet would eventually engulf Hudson Bay and Foxe Basin

and push westward for thousands of kilometres, terminating against the flanks of the Rocky Mountains. This grand model ultimately culminated in a continental-scale Laurentide Ice Sheet, with its centre more than four thousand metres thick. Flint envisaged the reverse of this gigantic process during the receding hemicycle of each successive ice age (the convention of the time was that there were four major ice ages, together with their "interglacials," forming the Pleistocene Period). The last remaining small ice caps and glaciers, therefore, were considered to have been located on the coastal mountains where the process began and where many exist today, as in the Torngat Mountains and Baffin Island. Flint's intriguing and highly persuasive thinking provided the Geographical Branch's Baffin Island project with the major objective of completing the initial challenge to Flint's conclusions that was initiated by the early work of the McGill Sub-Arctic Research Lab in Labrador-Ungava (Ives, 1957, 1960, 2010; Løken, 1960, 1962).

To this point, the Laurentide Ice Sheet, at its maximum, was regarded as a single immense dome, centred over Hudson Bay. Its southern perimeter extended deep into the United States; its western limit pushed up against the flanks of the Rocky Mountains, where it came into contact with the Cordilleran Ice Sheet complex; and its northern section engulfed Ellesmere Island and the High Arctic, with the exception of several of the northwestern islands, parts of the Yukon, and northern Alaska that were assumed to have remained ice-free beyond its limits. This would have placed the eastern margin of the Laurentide Ice Sheet along the edge of the North Atlantic/Baffin Bay continental shelf.

In 1960, therefore, the dominance of the "Flintian" hypothesis was beginning to be challenged. Nevertheless, the unknown proportion of the region's glacial history still vastly exceeded that of the known. Bruce Craig and John Fyles (1960) of the GSC had produced an overview of the history of the Laurentide Ice Sheet and argued that its northern limit was much more circumscribed than Flint had assumed.

They postulated an independent, or semi-independent, Ellesmere-Baffin ice complex, although they emphasized the sparseness of field data over a largely inaccessible landmass that was subcontinental in size. This immense region, of course, included Baffin Island and the uncertainty was a major factor in my early fieldwork aspirations.[6]

Arctic glaciology

In the years following the Second World War, Professor J. Tuzo Wilson worked to promote glaciological research in Canada. During the organization of the 1957 meetings of the International Union of Geodesy and Geophysics, which Canada was hosting in Toronto, Wilson urged the addition of glaciology as the one International Geophysical Year (IGY, 1957–1958) science not hitherto included by Canada. He also undertook to ensure a Canadian glaciology contribution to the IGY, which led to a number of initiatives: the first "Glacial Map of Canada"; a multi-year glaciological study of the Salmon Glacier in British Columbia; and many significant spin-offs—in particular, Geoff Hattersley-Smith's organization of Operation Hazen through the Defence Research Board, Ottawa, and his reconnaissance with Bob Christy (of the GSC) of the ice shelves along the northern coast of Ellesmere Island.[7] Separately, the Arctic Institute of North America (AINA) pioneered glaciological research on Baffin Island with the expeditions of 1950 and 1953 led by Pat Baird. Sometime later, Fritz Müller initiated the long series of McGill-Jacobsen expeditions to Axel Heiberg Island (Adams, 2007). Similarly, in the 1970s, Roy (Fritz) Koerner began a long-term glaciological project on the Devon Island ice cap with the Polar Continental Shelf Project.[8] Still, systematic and continuous glaciological investigations were non-existent in 1960—and did not start until the establishment of the Glaciology Section in the Geographical Branch. Significantly, the struggle to establish the Glaciology

Section received vital assistance from Tuzo Wilson (see chapter 3).

Some important new glaciological concepts were being proposed during this early period: for example, that all the previous winter's snowfall, even on the highest parts of the Barnes Ice Cap, melted during the following ablation season ("summer"), and nourishment depended on the refreezing of the meltwater (superimposed ice) onto the underlying very cold ice; also, that the Barnes Ice Cap could be a surviving remnant of the last ice age that had been substantially reduced in size during the intervening Climatic Optimum. The so-called end of the last ice age (Wisconsin) had been arbitrarily set worldwide at 10,000 years BP (i.e., at the beginning of the Holocene), leaving Baffin Island and many other glacierized areas as anomalies because of the persistence into the twentieth century of a large number of their glaciers and ice caps.

These were some of the intriguing issues that had prompted Geographical Branch interest—especially the need for verification, so that such major conclusions were not to remain based on a single season's observations or on hypotheses with very limited supporting field evidence. The late 1950s and 1960s were also the period when researchers began to realize that glaciers and ice caps in very cold climates were probably frozen to their beds. While no deep ice drilling had been undertaken at that time, it was hypothesized that actual glacial erosion would be severely restricted except under thick ice—that is, more than three hundred metres—the base of which would remain at the pressure-melting point and so allow the ice to flow over and to erode the bedrock. Similarly, the general understanding of the extent and depth of permafrost and associated landforms was slowly developing (Mackay, 1960, 1965; Brown, 1967). This was also a period when glaciology, as a scientific undertaking, was beginning to attract the attention of physicists, mathematicians, and engineers; thus, contributions to the "physics of ice" were emerging and affecting the hitherto much less rigorous "geographical" study of glaciers (Paterson, 1969). However, these early attempts to establish a systematic glaciological and glacier mapping program in Canada faltered; they were taken up again after 1970.

Botany

Prior to the establishment of the DEW Line, access to Baffin Island was almost entirely confined to visits to the small Inuit coastal settlements, aided by the Hudson's Bay Company annual supply vessels, occasional government icebreakers, and overwintering parties of the late nineteenth and early twentieth centuries such as those led by Bernhard Hantzsch, Franz Boas, Therkel Mathiassen, Dewey Soper, and Tom Manning. Plant collections had been made at many points and incorporated into regional assessments by such leading botanists as Erik Hultén, Erling Porsild, and Nicholas Polunin. The 1950 and 1953 AINA expeditions included botanists Pierre Dansereau and Fritz Schwarzenbach, as well as zoologists V. C. Wynne-Edwards and Adam Watson, so the second phase of more detailed and interdisciplinary investigation had begun in the Clyde and Pangnirtung Pass areas. Nevertheless, in 1960, Baffin Island was 95 percent unknown biologically, except by extrapolation from a few scattered points. This was especially true for our expedition areas, as is patently evident from the famous dot maps in Porsild's masterful 1957 work on the flora of the Arctic, which shows all locations of known vascular plant collections to that point in time. The scientific attraction of this virtual void was augmented by the provisional identification on the new RCAF trimetrogon air photographs of the extensive distribution of light- and dark-toned areas across the Baffin interior north and east of the Barnes Ice Cap. This raised the question that the tonal variation might be a manifestation of vegetation distribution and ground cover. Detailed studies of the vascular plants of northern Labrador and northeastern Baffin Island, comparable with those in Scandinavia, have not yet even been attempted.

Former high sea levels

Somewhat akin to the sparse knowledge of the distribution of plants was the very scattered information of field evidence for former high sea levels. Some information was certainly available—for areas along the northeast coast of Hudson Strait, the northwestern Foxe Basin, and areas along the northeast coast of Baffin Island. Nevertheless, systematic knowledge was entirely lacking and radiocarbon dating (C_{14}) was still unavailable. This lack of information was rendered more problematic because of the confusion caused by Flint's misinterpretation of Wordie's (1938) account of very high terraces in the inner fiords above 300 metres (see chapter 1) and Mercer's (1956) assumption of former sea levels in Frobisher Bay in excess of 340 metres above present sea level. Sim (1961) had made a reconnaissance of parts of Melville Peninsula for the Geographical Branch; however, with the exception of Løken's work in the Torngat Mountains, there had been no identification of actual strandlines nor their subsequent delevelling (tilting) that would record the direction of maximum postglacial rebound (related to former maximum ice sheet thickness). The scarcity of observations on former raised shorelines and significant misinterpretations attributed to high-level terraces were especially critical because of the widely understood relationship between maximum former sea levels and thickness of the ice age ice sheets.

The scientific objectives

When Dr. Norman Nicholson sought to recruit an experienced physical geographer in 1959–1960, it is understandable that the various topics outlined above ensured that Baffin Island would offer a remarkable field research opportunity. Linked together, they were an essential element of my interest in joining the staff of the Geographical Branch. While by no means unique for Canada's Arctic and Subarctic, Baffin

Island stood out as an obviously attractive place for expansion of the research emanating from the McGill Sub-Arctic Research Laboratory in Labrador-Ungava despite the logistical challenge it posed. These considerations, therefore, provided the basis for the 1961 reconnaissance to Rimrock, Flitaway, and Separation lakes and the trans-island traverse south of the Barnes Ice Cap (chapter 2). And despite the inadequacies of the aircraft charter, the reconnaissance was sufficiently successful that it led to a continually expanding research program. The main results are highlighted in the following sections.

The 1961 reconnaissance

A virtual jigsaw puzzle—that of the light- and dark-toned expanses, the abandoned glacial lake shorelines, and the multiple ridges running perpendicular to the trend of the main valleys north of the Barnes Ice Cap—was provisionally resolved (Figs. 2, 3, and 9). The light-toned areas resulted from the very limited growth and small diameter of rock lichens that left patches of land nearly barren, in contrast with the darker intervening areas whose lichen cover was heavy. It was concluded that this was the result of differential distribution of permanent ice and snow at some time in the past that had either killed off a former mature lichen cover or had inhibited lichen growth in comparison with the areas that remained ice-free. This period of former, more extensive ice cover was originally estimated to have occurred some two to four hundred years ago, based on extrapolation of lichen growth rates. The actual dating estimate needed to be refined and applied more systematically to the area below the pronounced shoreline of a former glacial lake (Glacial Lake Lewis) that similarly marked an upper limit of diminished lichen cover. Nevertheless, it was intuitively compelling to attribute these areas of limited lichen cover to the period of the Little Ice Age (AD 1500–1900), when snowline lowering placed north-central Baffin Island

on the brink of instantaneous glacierization. This strengthened the earlier hypothesis of rapid ice age initiation and growth across the Labrador-Ungava plateau (Ives, 1957, 1960b, 1978). And as the main glacial lake shoreline proved to be horizontal (unlike the much older deglacial shorelines, which were tilted), additional support was provided for the assumption that the period of imminent glacierization was recent (Ives, 1962). This assumption was extended by Falconer's subsequent investigation of the Tiger Ice Cap (in northern Baffin Island), the retreat of which was exposing masses of apparently dead plant material (lichens and mosses). Falconer (1966, p. 198) raised the possibility that moss spores may have survived a long period of glacial entombment and that "it cannot be safely assumed that they [the rock lichens and mosses] are dead." At a much later date, these early findings were seen as apparent proof of very-long-term survival (i.e., more than fifty thousand years) of plants beneath thin ice patches presumed to have been frozen to their beds (G. Miller, letter to G. Falconer, November 22, 2013; La Farge et al., 2013; Miller et al., 2013).

The early attempts to interpret the intriguing arrangement of the light- and dark-toned areas quickly went beyond the simple identification of probable partial cover of semi-permanent ice patches that likely existed two to four centuries ago (Beschel, 1957, 1961, Ives, 1962). However, these early assumptions were based on limited evidence and eventually met with strong opposition. Koerner's (1980) paper raised substantial objections to the Baffin project interpretation and caused a significant pause in the scientific exploitation of this aspect of lichenometric application. Much later, more detailed research eliminated Koerner's challenge (Wolken, England, & Dyke, 2005), leaving the original interpretation intact.

The multiple trans-valley ridges were identified as subglacially formed moraines that had been squeezed up into basal crevasses, close behind the calving ice front of earlier extensions of the proto–Barnes Ice Cap that had impounded a series of ice-dammed lakes.

This was initially proposed as a hypothesis in 1961 (Andrews, 1963), although it was later substantiated by additional research (Andrews & Smithson, 1966).

Shell collections by Sim and Ives from former raised marine shorelines along the west coast of Baffin Island were combined to produce the first marine uplift curve for Foxe Basin. There was associated evidence to indicate that an ice divide, centred over Foxe Basin, had been displaced northeastward onto Baffin Island prior to about five thousand years ago, although outlet glaciers from the proto–Barnes Ice Cap were still in contact with the sea (Ives, 1964; Sim, 1964; Andrews, 1966, 1970; Dyke et al., 2002; Miller et al., 2002). A general outline of the glacial history of the northeastern Canadian Arctic, with a semi-independent ice sheet centred over Foxe Basin, was developed (Ives & Andrews, 1963). The proposed "Foxe Dome" was especially significant as it marked the first persistent formal departure from the single-domed Flintian model of the Laurentide Ice Sheet (centred over Hudson Bay), which had been suggested much earlier by Joseph Tyrell of the Geological Survey of Canada (1898) but had been largely abandoned in the literature in favour of Flint's reconstruction. Finally, the overall review paper stemming from the 1961 reconnaissance (Ives & Andrews, 1963) emphasized the importance of the Cockburn Moraines.

The results from the 1961 reconnaissance and the overall discussion of the earlier prevailing status of the several aspects of thinking about ice age history of the Canadian Arctic laid the foundation for progressive enlargement of our research objectives for the following six years and far beyond.

Highlights of the 1962–1967 research

By the close of the 1962 field season, it had been demonstrated that the highest marine shore features along parts of the northeast coast fronting Baffin Bay did not exceed about eighty metres above present sea level. In addition, after the long outlet glaciers of the last ice age that had flowed down the fiords into Baffin Bay and across the continental shelf had retreated, local mountain ice caps and glaciers had expanded and cut through the older lateral moraines and reached tidewater in the fiords. It could be seen that the extensive fiord outlet glaciers had deposited long lateral moraine systems that were related to the Cockburn Moraines, although the precise relationships and timing still needed to be determined. However, collections of seashells from raised marine terraces that intermingled with the lateral moraines indicated that some of the fiord heads retained calving glaciers until about seven thousand years ago. Even at this early stage of the project, considering also the preceding work in the Torngat Mountains, it could be claimed that Flint's general thesis (especially that ice retreat had occurred much earlier) was no longer tenable.

Thereafter, the field results become so numerous and complex that only the highlights are summarized here:

Mass balance and internal motion of the Barnes Ice Cap

Sagar and Løken determined that the Barnes Ice Cap was experiencing an annual negative mass balance more often than not, although the southwest side was more negative than the northeast side. In effect, the Barnes Ice Cap was still being displaced slowly toward the northeast, as it had been for thousands of years since its centre had been situated over Foxe Basin. Furthermore, extensive traverses across the length and breadth of the ice cap by Housi Weber provided seismic and radio-echo sound data on its thickness and on the nature of the subglacial topography. This confirmed the earlier hypothesis that the Barnes Ice Cap is a relic of the last ice age. Løken also demonstrated that the southwest margin had experienced a significant readvance, possibly a surge, sometime in the recent past (Sagar, 1966; Løken & Sagar, 1968).

Glaciological and hydrological innovations

Østrem introduced new methods for studying the hydrology of turbulent glacier meltwater rivers (Church & Gilbert, 1975; Østrem et al., 1967) and initiated Canada's contribution to the glaciological objectives of the International Hydrological Decade with mass balance studies of the specifically named Decade Glacier (Inugsuin Fiord) and the Lewis Glacier (northwestern Barnes Ice Cap). Equally important was his initiation of a glacier mass balance transect across the Rocky Mountains and British Columbia Coast Ranges (Østrem, 1966); along with Müller's research on Axel Heiberg Island, this was one of the earliest Canadian long-term initiatives, followed later by Koerner's work on the Devon Island ice cap. While long-term study of the five glaciers selected for the transect did not survive the dissolution of the Geographical Branch, work on the Peyto and Place glaciers has continued to the present, rendering that the longest continuous record of mass balance on Canadian glaciers.[9] With Falconer (1962), Østrem also set in motion a series of publications intended as a complete inventory of Canadian glaciers (Falconer, Henoch & Østrem, 1966). Post-1967 work on a glacier inventory for Canada, while extensive, has been intermittent. An early version was produced by the Inland Waters Branch (Henoch & Stanley, 1970). An outstanding and comprehensive account of the history of glacier research in Canada has been provided by Ommanney (2005).

The significance of very old marine molluscs

Løken (1966) collected mollusc shells (seashells) from raised coastal deltas along the northeast coast (Cape Aston and Clyde Foreland) that were radiocarbon-dated at older than 54,000 years BP. This was a first for the Canadian Arctic, initially strengthening support for the nunatak hypothesis—that not only mountaintops but sections of the coastal lowlands remained ice-free for at least the period of the last ice age maximum (Wisconsin). This interpretation

would have provided for a variety of ice-free habitats, or refuges, for arctic-alpine plant species. The mountaintop evidence is discussed below. Although this conclusion has since been strongly refuted (Dyke et al., 2002; Miller et al., 2002; Sugden & Watts, 1977), a recent publication by Miller et al. (2013) appears to reverse the refutation. This recent work identifies thin ice patches and glaciers that persisted on the interior highlands north of the Barnes Ice Cap, from the retreating margins of which plant material has been collected and dated possibly as far back as the pre-Wisconsin interglacial (120,000 years ago).[10] My personal interpretation is that such thin ice patches retaining their independence of the Laurentide Ice Sheet at its maximum would place a severe restraint on conclusions concerning its size. While this does not eliminate the possibility that the Clyde and Cape Aston forelands were not themselves mantled with thin inert ice frozen to its bed during the maximum of the last ice age, it appears to reopen the debate. Further to this, recent work from Oslo (Brochmann et al., 2003) has also reintroduced the likelihood that at least a small number of vascular plants of the Amphi-Atlantic group survived the maximum of the last ice age (*Weichselian*, a European term) on *nunataks* in Norway (this controversial hypothesis is elaborated further below). Løken made extensive collections of mollusc shells and examined the sedimentary stratigraphy of the low sea cliffs extending northwest from Cape Christian (Clyde Inlet). This resulted in the invitation to Dr. Rolf Feyling-Hanssen, a marine palaeontologist, to undertake an intensive study of this thirty kilometres of cliff exposure. Rolf's research led to the identification of the area as one of the most important glacial/interglacial (and possibly pre-Pleistocene) stratigraphic sites in the Canadian Arctic (Feyling-Hanssen, 1967, 1976).

Large-scale mapping of crustal (glacio-isostatic) rebound and sea level change

Extensive studies by Andrews of the relation between glacial and raised marine landforms along both the eastern (Home Bay) and western (Foxe Basin) coasts of Baffin Island were extended into Hudson Bay (Andrews & Falconer, 1969). From this research, Andrews produced a leading monograph on late-glacial and postglacial sea level changes, recording the crustal unloading (rebound) induced by the regional retreat of the northeastern sector of the Laurentide Ice Sheet across the eastern Canadian Arctic (Andrews, 1970). As a result of this monograph, Andrews was presented with the prestigious Kirk Bryan Award of the Geological Society of America in 1973. The work also confirmed the initial 1961 proposal (Ives & Andrews, 1963) that Foxe Basin had been a principal centre of ice dispersal for the northeastern sector of the Laurentide Ice Sheet during the maximum of the last ice age, as indicated by the tilt of shoreline features created by former high sea levels. However, Andrews's contention that the marine limit determinations in northern Foxe Basin depicted a contemporaneous surface was later modified. The marine limit there was formed during sequential ice margin retreat, as was the case in most other places (A. Dyke, personal communication, June 26, 2014). Subsequently, ice flow out from the centre of Foxe Basin northeastward across Baffin Island was reversed as the ice divide migrated toward its present position along the crest of the Barnes Ice Cap.

Use of helicopter to test the nunatak hypothesis

Using a helicopter to access many of the high mountaintops between Ekalugad and Gibbs fiords, and several of the outermost forelands between the fiords (e.g., Cape Aston, Cape Christian, Cape Henry Kater, and Remote Lake), led to a much fuller understanding of ice conditions at the maximum of the last ice age. As in the Torngat Mountains of northern Labrador, a distinct upper limit (the glacial trimline) to indisputable glacial erosion and deposition was traced along extensive sections of the fiords from higher than seven hundred metres asl near the fiord heads to below sea level on the outer coast (that is, most of the fiords

between Home Bay and North Arm, just south of Pond Inlet). The glacial evidence was principally in the form of long stretches of glacial lateral moraines below which evidence of glacial erosion was profuse. But above them, where mountain summits were gently sloping or plateau-like, the surface was mantled by a deep cover of angular frost-shattered bedrock, from which occasional tors projected. This type of surface, in both Norway and the eastern Canadian Arctic, was widely accepted at the time as evidence that the upper levels had not been overridden by erosive masses of ice at the glacial maximum. Nevertheless, occasional glacial erratics of unknown age were detected on the high summits among the tors and mountaintop detritus (see Fig. 32). This work was followed up by a long series of field investigations much farther south, in the vicinity of Broughton Island and the Penny Highlands, that was organized after 1967 by John Andrews from INSTAAR.

Due to the lack of methods in the 1960s for actual dating of rock surfaces (either tors or erratics), it was concluded that at some time in the past the continental ice sheet had overtopped the highest summits. However, the absence of observable indications of glacial erosion on these surfaces indicated that the ice must have been sufficiently thin and cold to remain frozen to its bed, leaving the boulder and tor surfaces essentially unaltered. In contrast, the ice flowing along the line of the fiords (i.e., outlet glaciers) would have been extremely thick and faster flowing, characteristic of warm-based ice, facilitating the extensive overdeepening of the fiords by glacial erosion. Løken's (1965, 1966) dating of coastal sediments as older than 54,000 years, supplemented by the work of Andrews, Buckley, King, England, and others, together with the dates on the Remote Lake outer moraines, led to the conclusion that any "ice-free" areas delimited by mountaintop detritus and tors predated the Last Glacial Maximum. It was presumed that the scattering of erratics must have been emplaced by a thin cover of ice during an early ice age. This thinking was reinforced by the observation that many of the higher and flatter mountain summits are today mantled by a thin carapace of ice, presumably frozen to the ground surface and from which angular boulder fields are emerging due to the overall northern hemisphere glacier shrinkage of the early twentieth century. However, no work was undertaken on the mountaintop flora; accordingly, the earth science–oriented research moved progressively away from the nunatak hypothesis *per se*. The later introduction of cosmogenic nuclide exposure dating and other techniques has led to significant revision of this rather simplistic interpretation (see below).

Following the Baffin Island expeditions of the 1960s, continuation of this research from INSTA-AR, in Boulder, Colorado, appeared to confirm the general hypothesis of a minimum extent and thickness of the Laurentide Ice Sheet during the last ice age, as outlined above. Numerous publications resulted, similar to those of the 1960s. However, they were much more detailed and included a major focus on identification of "weathering zones"; the highest zones, which included the mountain summits and high plateaus, were usually identified as likely to have remained above the maximum limits of the main ice sheet (see also Ives, 1960b). Nevertheless, the occasional enigmatic glacial erratic located on and among tors and weathering pits and within the mountaintop detritus was recorded. Thus, ambivalence persisted until the advent of cosmogenic nuclide exposure dating, as mentioned above, which seemed to provide definitive ages for both the tors and the erratics. A general conclusion was reached: that at the maximum of the last ice age, the Laurentide Ice Sheet was much thicker than previously presumed, had overtopped all the highest summits, and had extended to the edge of the continental shelf.

Regardless, the enigma resurfaced. Art Dyke (pers comm., 15 February 2014) kindly sent me the abstract of a paper to be presented during the 2014 General Assembly of the European Geophysical Union (Margreth et al., 2014). In brief, the research on which the paper is based demonstrates serious inadequacies in the hitherto widely accepted accuracy of the earlier cosmogenic nuclide exposure dating techniques. The authors argue that many of the Baffin Island tors were not covered by ice throughout the last ice age, and some may have been exposed continuously for much longer. It is fascinating that this latest explanation of the enigmatic tors/erratics/mountaintops and maximum thickness of the Laurentide Ice Sheet, while based on many times more research and much more refined methods than were available in the 1960s, remains essentially unchanged (in other words, at least a partial double reversal of thinking has taken place between the 1960s and the present). To my mind, however, there still is no adequate explanation for the presence of glacial erratics on top of tors if the summits were covered only by thin, scarcely moving, or immobile ice patches or ice sheets frozen to their beds.

Final collapse of the Laurentide Ice Sheet

The extensive study of raised marine shorelines and their dating by radiocarbon determinations, together with the systematic mapping of the extent of the Cockburn Moraines, caused us to extend our review of available reseach results and to consider the moraine systems along Melville Peninsula (Sim, 1961) and the Arctic mainland coast. These considerations prompted Falconer to introduce his copy of the draft glacial map manuscript on which he had worked under the direction of Tuzo Wilson. This showed extensive moraine systems that he had sketched but which had not been included on the final printed map (Wilson et al. 1958). In addition, contemporaneous research along the Arctic mainland coast greatly strengthened the early air photo interpretation taken from Falconer's draft manuscript (Lee, 1959; Blake, 1963). From this, it was not excessively conjectural to propose continent-wide projections through Keewatin and northern Ontario/Quebec to produce a theoretical depiction of the final stage of the Laurentide Ice Sheet as it existed approximately eight thousand years ago. And from this, it was concluded that the next phase of late-glacial Laurentide Ice Sheet retreat culminated with the catastrophic collapse of its geographic centre in Hudson Bay as Atlantic waters rushed in to disrupt it. This left remnant ice masses centred on Labrador-Ungava, interior Keewatin, and Baffin Island–Foxe Basin (Falconer, Andrews, & Ives, 1965; Falconer, Ives, Løken, & Andrews, 1965). This concept was revolutionary for the time (i.e., the mid-1960s) and was hotly contested, although the general interpretation has subsequently received convincing support. Viewed from the perspective of the massive discharge of the gigantic Glacial Lake Agassiz, the 1965 papers have received substantial confirmation. According to the much more recent research, Glacial Lake Agassiz broke through its Laurentide Ice Sheet dam to flow into the North Atlantic via Hudson Bay and Hudson Strait. The recent dating of this event at 8,200 years ago (Barber et al., 1999; Clarke, Leverington, Teller, & Dyke, 2004; and many others) coincides remarkably with our earlier prediction of approximately 8,000 years ago, viewed from the opposite side (i.e., the northeast side) of the remnant Laurentide Ice Sheet. As implied above, over the last several decades, knowledge of the deglaciation of North America has advanced prodigiously. One of the most outstanding contributions has been made by Art Dyke (2004), closely matched by John England and Gifford Miller based on decades of fieldwork and numerous earlier publications. These contributions can also be regarded as outgrowths of the 1961–1967 expeditions to Baffin Island.

Lichenometry

Under the direction of Roland Beschel, the initiator of lichenometry, Andrews and Webber (1964, 1969) built on the method Beschel had developed in Austria and West Greenland (Beschel, 1957, 1961). They focused only on epipetric lichens. During the 1963 field season, Webber and Andrews, assisted by a virtual army of students, painstakingly produced thousands of measurements on lichen diameters and percentage of lichen cover at varying distances from the Barnes Ice Cap margins, especially in the area stretching from Flitaway Lake southwestward to the confluence of the Isortoq, Striding, and Lewis rivers. Measurements were made on a number of different species so that multiple growth curves could be constructed and cross-checked for their value in determining the age of various periglacial and glacial features comprising boulder and rock surfaces. A number of lichen stations were established where individual lichens were photographed and their outlines traced on mylar sheets. The establishment of these stations was a response to Beschel's prescient urging. Rates of lichen colonization and growth were estimated using reference points of age-since-deglaciation. This was made possible by the combination of excellent air photographs from 1948 and 1961, counts of growth rings of willow taproots, and parsimony about the likely time required for stabilization of the reference points. The two most useful lichens were those of the very slow-growing crustose, yellow-green *Rhizocarpon geographicum* group and the faster-growing foliose-fruticose black *Pseudephebe (Alectoria) miniscula*. From both the data sets and Beschel's experience in West Greenland, Andrews and Webber (1964) reasoned that both lichens grew at a more or less constant rate across the study area and that each had straight-line growth curves. They estimated the diameter expansion for *Rh. geographicum* to be 0.064 millimetres per year and for *Ps. miniscula* to be 0.4 millimetres per year. The largest *Rh. geographicum* thalli were about 50 millimetres and the largest *Ps. miniscula* were about 130 millimetres. The former clearly would be on the order of one thousand years in age. The latter would be getting quite senescent by 130 millimetres and thus provided only a limited age range, but they were particularly useful for Little Ice Age features up to about three hundred years in age. From these rates and from the distribution of maximum diameters across the study area, maps were drawn and published giving the outlines of former glaciers and ice-dammed lakes.

Later research around the southern margins of the Barnes Ice Cap and, farther south, on Cumberland Peninsula has shown that the earlier work stands up extremely well (Andrews & Barnett, 1979; Barnett, 1977; Miller 1973a, 1973b). This work included a revisit, after forty-six years, that provided individual growth rates for ninety-five lichen thalli, thus re-affirming the importance of Beschel's urging. A new reference growth curve with 95 percent confidence limits became available. One specific result of this later work is that, at the northwestern end of the Barnes Ice Cap, the extended and combined small Lewis and Pintail glaciers damming the last remnant of Glacial Lake Lewis at its 268-metres-asl shoreline began to retreat in AD 1788 (give or take thirty years). This caused total drainage of the lake. Developing out of the burst of Baffin Island lichen studies from 1963 to 1979, *A Manual on Lichenometry* (Locke, Andrews, & Webber, 1979) was published and has since expanded worldwide to become the prime reference on the subject.

Still, recent work on the challenges of perfecting lichometry over the last two decades has revealed extensive complications. The assumptions about the feasibility of the simple transfer of lichen growth rates from one region to another and between different rock types have been seriously disputed. Regardless, the comparability of sites within the area of north-central Baffin Island seems basically defensible.

[Author's note: The foregoing section was generously revised and expanded by Pat Webber (personal communication, February 20, 2014).]

Plant ecology and plant collections

Webber's 1963 and 1964 work investigated the nature of the structure and organization of plant communities in the greater Lewis River area; it was founded on extensive collections of vascular plants, mosses, and lichens and the establishment of eighty-nine one-by-ten-metre quadrats. While the flora as a whole are not especially rich (only 84 vascular species around the Lewis Glacier, with 142 on the warmer, older west coast around Eqe Bay), Webber collected and identified more than 1,300 specimens, which were later generously verified by experts at the National Museum of Natural Sciences (today, the Canadian Museum of Nature): A. E. Porsild, I. M. Brodo, and H. A. Crum. The moss collections were re-examined by Guy Brassard and Allan Fife (Brassard, Fife, & Webber, 1979). These thorough collections filled the void mentioned earlier. Most specimens were collected in at least triplicate, meticulously documented by place of collection, labelled by Webber, and placed in world-class herbaria. The first set also went to the Canadian Museum of Nature, where it is now integrated into the main collection and is kept current with regard to nomenclatural changes; Pat has reported (personal communication, March 10, 2013) that the specimens have frequently been annotated by visiting specialists. Replicate sets of the vascular plants can be found in the Fowler Herbarium at Queen's University and the William A. Weber Collection at the University of Colorado Herbarium. The quadrat data and plot photos are also well archived and digitized; the plot data will be held in the Arctic Vegetation Archive (see Villarreal et al., 2014; Walker, 2014; Webber, 2014).

In 1962, John Andrews and Bruce Smithson discovered extensive plant-bearing beds on the east bank of the Isortoq River some ten kilometres below its confluence with the Lewis River. The deposits were sampled extensively in 1963 along with a similar, probably contemporary, isolated mound of deposits at Flitaway Lake discovered by graduate student Dave Harrison. The palynology and plant macrofossils were described by Terasmae et al. (1966). The macrofossils dated beyond the realm of radiocarbon technology and were assigned to the Sangamon Interglacial. The deposits contained a number of well-preserved remains of plants no longer present in today's local flora, for example, *Ledum groenlandicum* (Labrador tea) and *Betula nana* (Dwarf birch). Modern pollen rain was collected for two years at the Flitaway climate station and used as reference. The modern flora and the careful analysis of the fossil material provided a clear picture of the ancient climate that would have been wetter by some thirteen centimetres per year and warmer in July by one to four degrees Celsius. The growing season was twenty to twenty-five days longer than the present ninety days. Re-examination of these deposits with the newest dating techniques, and their use as a yardstick given the present warming climate, should provide worthwhile scientific rewards. The work also pointed to the value of *Betula nana* as a phytogeographic and climate zonal indicator. For example, Jacobs, Herm, and Luther (1993), in follow-up pollen analysis in southern Baffin, referenced the presence of *Betula nana*—a short but erect shrub presently restricted to southern Baffin and a good indicator of the Low Arctic. It is an amphi-Atlantic plant that figures prominently in the recent, and best, whole Arctic vegetation map (CAVM team, 2003).

In his doctoral dissertation, Webber (1971) established the paradigm that, while High Arctic vegetation was best viewed as a continuum classification into reasonably discrete entities, it could go hand in hand with ordination. Ordination analysis was the best way to correlate plant distribution with environmental controls, and classification was the best basis for communication, mapping, and experimentation where replication was needed. Such notions will be seen today as "old hat," but there was much controversy in the 1960s about the levels of organization within the Arctic plant community.

As Pat noted, it was somewhat serendipitous that the plant collections, Quaternary paleobotanical studies, and quadrat data now form a benchmark and a useful basis for assessing the consequences of a warming Arctic.

The 2009 Back to the Future project (Callaghan, Tweedie, & Webber, 2011) under the auspices of the Fourth IPY was able to resample and precisely geo-reference eighty-seven of the Webber quadrats. Two student participants, Mark Lara and Sandra Villarreal, who represent the latest generation of Arctic botanists, included in their doctoral dissertations (Lara, 2011; Villarreal, 2012; Villarreal et al., 2014) the new lichenometry picture and the recorded vegetation changes. They were able to take advantage of John Jacobs's newly assembled sixty-year climate record for Central Baffin Island and the Lewis and Flitaway areas, which shows a sustained increase in July temperatures and summer warming index (SWI) since the mid-1960s. Lara and Villarreal demonstrated that the Lewis Valley is greening (Bhatt et al., 2010) and that while the communities are changing, their trajectories and explanation of these changes are complicated because of the interaction of warming and ageing of the surfaces (that is, retreat of the glacier and ice cap margins). Lara focused on community functional changes such as light reflectivity and productivity of the vegetation over a decadal timeframe using gas exchange, soil moisture, and spectral reflection methods. Such methods had been only a pipe dream in the 1960s. Villarreal showed, using an updated classification and ordination, that the number of plant species, primary productivity rates, and extent of ground cover have significantly increased since the 1960s. The greatest change was noted in plant communities with high soil moisture. For example, two pond margin communities, *Campylium-Aulacomnium*-moss meadows and *Eriophorum-Pleuropogon* wetlands, had increased in biomass by 178 percent and 46 percent, respectively, while greenness had increased 35 percent and 16 percent, respectively. Soil moisture was found to have decreased in *Carex* stand wet meadows and *Campylium-Aulacomnium*-moss meadows by 30 percent and 24 percent, respectively. Overall, Villarreal and Lara found evidence of a general drying of the landscape and of dramatic changes close to the ice cap, suggesting that plant community succession (vegetation cover and species richness) has accelerated over the past half century—especially on surfaces that have been exposed from the ice for less than two hundred years.

The 2009 IPY crew noted the increased frequency of *Astragalus alpina* (Milk vetch), *Vaccinium uliginosum* (Blueberry), and *Salix Richardsonii* (Erect willow), which had previously been rare to the area. Only one solitary blueberry plant was seen by Webber during all his perambulations around the Lewis Glacier in 1963 and 1964; yet the species was common enough near the Isortoq plant fossil beds in 2009, where Gifford Miller reported that he was able to add handfuls of blueberries to his breakfast cereal. This certainly could not have been done by John Andrews and Pat Webber in the 1960s! The increased stature of *Salix Richardsonii* is commensurate with the general Arctic-wide increase of erect shrubs reported by Tape, Sturm, & Racine (2006).

These findings demonstrate the importance of careful records and a need to continue to monitor tundra landscapes over decadal time scales in our changing world.

[Author's note: The foregoing section was generously contributed by Pat Webber (personal communication, February 28, 2014).]

Submarine topography

Løken's shipboard submarine surveys provided extensive information on the topography of the continental shelf off the eastern Baffin Island coast and along many of the fiords (Løken & Hodgson, 1971). This information was vital in its own right for improving navigation safety. It also inaugurated what has become a central issue concerning the marine geology not only of the Canadian Arctic but also of other glaciated continental shelves in both the Arctic and Antarctic. Further, Løken's marine research provided an essential complement to the longstanding emphasis of the time spent on the terrestrial record on adjacent islands, such as Baffin Island and Greenland.

General influence on subsequent research

As inferred earlier, for many years following 1967, Baffin Island constituted the primary target of the developing Arctic element of the research activities of INSTAAR under the leadership of John Andrews. In fact, I was able to initiate and host an annual "Baffin workshop" for inter-institutional meetings in Boulder, Colorado; the workshop subsequently became the International Arctic Workshop under the guidance of John Andrews. It is held in alternate years in Boulder and at other leading research centres in the United States, Canada, Norway, Iceland, Sweden, and Denmark. It has continued without interruption to the present and, whereas it began informally with only a handful of faculty and graduate students, it is now attended by hundreds.

The total amount of research on Baffin Island has multiplied to many times that carried out in the 1960s. It is far too extensive to detail here; also it would go far beyond the scope of this book and the competence of the author. However, it is reasonable to claim that the Geographical Branch endeavour provided a critical stimulus and seeded many internationally acclaimed careers in Arctic environmental science, many still active in numerous university and governmental agencies around the globe. Indeed, many of these careers are now at least three "academic generations" removed from the Canadian Geographical Branch and its initial beneficiary, INSTAAR. The prospect of compiling the many hundreds of available research papers and publishing a major book (or books) from them would be a formidable task. Yet until that is done, we will not realize the full value and impact of the now more than a half century of Arctic field research.

GLACIERS
AND GLACIAL
LANDFORMS

All photographs were taken in the 1960s,
so remarks such as "in recent decades" refer
to the time period prior to the 1960s.

Fig. 83: Giant boulder forms summit of high mountain. This huge boulder forms the "capstone" of a 1,200-metre mountain in outer Inugsuin Fiord. Note human figure for scale. There are no indications in the immediate vicinity of former glacier activity, neither erosional nor depositional. Interpretation becomes a problem. Is it an erosion residual, such as a more resistant part of the bedrock left standing after eons of weathering? Or was it emplaced by a former glacier or continental ice sheet? Without supporting evidence, no definite conclusion can be drawn. However, a similar boulder forming the summit of Anvil Mountain more than sixty kilometres to the southwest supports a telltale small glacial erratic on its surface (see Fig. 32). It was therefore concluded that both boulders were probably put in place by moving glacier ice.

Fig. 84: A pseudo-erratic immediately before "birth." The complex Precambrian bedrock of the eastern section of
the Canadian Shield contains numerous rock types that differ markedly from the country rock into which they were
incorporated. Differential weathering over a long period of geological time can release them onto the surface amid the rubble
of mountaintop detritus. The example shown here is not yet detached. Some distant future researcher may discover it loose
on the surface and conclude (incorrectly) that it is a glacial erratic. Examples were found in the same vicinity in all stages of
"parturition." We were even able to break one loose. This is a warning against hasty interpretation.

Fig. 85: Weathering pits constitute another possibly false lead. This example cannot be passed off as the ancient footprint of a polar bear. Large numbers of weathering pits occur in the summit zone of the high coastal mountains of Baffin Island and Labrador. Usually found within the mountaintop detritus, they have been interpreted as evidence of non-glaciation—or at least for the bedrock not having been scoured by actively moving ice. However, examples of weathering pits have been found at lower levels, although in much fewer numbers. These may have been protected in some way, for instance, under a cover of till. Again, interpretation is controversial, although statistical analysis has suggested that their numerically great preponderance in the summit zone may indicate an absence of actively eroding ice or a thin, cold cover of ice frozen to the surface.

Fig. 86: Positive evidence of former glacial erosion. In contrast to the upper levels, or summit zone, glacially smoothed bedrock is both conspicuous and widespread at the lower levels. This example includes glacially moulded bedrock with a partially developed *roche mountonée* form. Perched boulders, even of the same rock type as the underlying bedrock, are definitely glacial erratics. The smaller cobbles, sand, and clay constitute glacial moraine (till). The moulded surfaces may show glacial striations, although they are also easily weathered by the coarse-grained bedrock.

Fig. 87: Glacier showing no indication of recent retreat. This glacier, located near the head of Itirbilung Fiord, is in close contact with extensively vegetated ground—mainly mature rock lichens and mosses. There is no evidence that any retreat had occurred in the preceding decades.

Fig. 88: Glacier showing evidence of extensive retreat. Little more than thirty-five kilometres from the glacier in Fig. 87, this glacier has obviously retreated more than two kilometres in recent decades. The very light tone indicates scant vegetation cover (i.e., lichens). The glacier's former extent is displayed by the massive end moraines and associated features. Located near the head of McBeth Fiord, its orientation (north-facing) and gradient are similar to those of the Itirbilung glacier. The two photographs clearly illustrate very different glacier responses in a comparable environment.

Fig. 89: Medial moraine formation, inner Clyde Inlet–Ayr Lake vicinity (indicated by arrow). The lateral moraines of two small converging glaciers coalesce as a medial moraine. Note also the way in which the crevasses have developed.

Fig. 90: Textbook example of a glacier and its marginal and frontal deposits. The outlet glacier descending from a highland ice cap north of inner Clyde Inlet displays a variety of associated landforms: the relatively recent (i.e., of the last 100 to 250 years) multiple-ridged lateral and end moraines; a conspicuous glacial outwash plain, or sandur, graded to sea level in the fiord; old, subdued end moraines on the outer edges of the sandur (marked by arrows); and medial moraines on the present glacier surface.

Fig. 91: The piedmont glacier. Where valley or outlet glaciers extend onto more gentle slopes—in this case, the floor of a wide pass—they often expand into a spoon shape under their own weight, hence the term "piedmont" (foot of mountain). The two piedmont glaciers shown here have blocked the pass, damming two lakes, the one on the left is still ice-bound.

Fig. 92: "Skeleton" of a glacier. In regions with extremely cold climates, as glaciers retreat, thin, and waste away, many of their deposits and associated landforms are preserved, sharply etched on the landscape. This example shows remarkably distinct multiple moraine ridges, glacial outwash plains, and moraine-dammed lakes. As the climate gradually ameliorates, such features become much less distinct and may even disappear. This is because they have ice cores, with ice forming as much as 90 percent of the entire feature.

Fig. 93: Emergence of multiple eskers from glacier terminus (inner Inugsuin Fiord). The retreating glacier terminus has revealed a startling display of perfectly preserved eskers more than forty metres in height. The preservation is due to their extensive ice cores; therefore, subsequent climate warming will likely cause their collapse or even their entire disappearance as recognizable eskers. It would be interesting to replicate this photograph today—forty-eight years after it was taken—to see the effects of recent climate change.

NOTES

Introduction

1 The McGill-Carnegie-Arctic Research Program had been established several years previously to help counter the shortage of scholars involved in Arctic research. My initial immigration to Canada was made possible by the award of one of the scholarships of that program.

2 Today the Glaciology Section is part of the GSC. The original design for the glaciology program included an east–west transect across the Rocky Mountains and the Coast Ranges for long-term glacier mass balance studies and a glacier inventory for the entire country in addition to the work on Baffin Island.

1 | Baffin Island: The Place and the Research

1 Inuktorfik Lake ("the place where they ate human flesh") was named after a tragic occasion when an entire Inuit hunting party was marooned by early snowmelt, all but one succumbing to starvation and traditional cannibalism. The name was printed on the topo map (scale 1:250 000) that we used in the 1960s. It is still there, as of 2013.

2 Recently Dr. Patricia Sutherland has uncovered archaeological evidence that indicates long-term residence by Vikings within Hudson Strait and possibly along the northeastern coast of Baffin Island (Sutherland, 2009).

3 Determination to prevent the disappearance of this name from Canadian cartography prompted a recommendation for the 1:500 000 map sheet name "Cockburn Land" and naming of the extensive system of glacial moraines as the Cockburn Moraines. Sir George Cockburn (pronounced "Coh-burn") was the first chairman of the Arctic Committee and Lord of the Admiralty, 1834–1835 and

1841–1846. W. E. Parry says he named the northern portion of "Baffin's Land after Lord Cockburn whose warm interest in everything relating to northern discovery can only be surpassed by the public zeal with which he always promoted it" (Parry, 1824, p. 330).

4 There undoubtedly were numerous Inuit place names, and likely multiple names for the same feature, possibly due to the lack of contact between different groups, although these were unknown to us at the time. More recently, many of the old Eurocentric names have been replaced with Inuit names. Perhaps the most prominent is the highly appropriate substitution of Iqaluit for Frobisher Bay.

5 Several of these place names, and others not listed here, should also be attributed to the Scottish whalers.

6 W. Vaughan Lewis was reader in geography, University of Cambridge, and one of the leading glacial geomorphologists of the period. He provided much advice and assistance for our undergraduate student expeditions to Iceland (1952–1954). He was on his way to visit me at the Geographical Branch in the early summer of 1961 when the car in which he was a passenger, en route to O'Hare Airport in Chicago, was struck by a gasoline tanker, causing his tragic death.

7 Cape Christian was a U.S. Coast Guard station operated from Boston. It used the LORAN (long range navigation) system and supported aviation and sea operations out of Thule, Greenland. Not a DEW Line station, it was strictly off limits.

8 Unfortunately, the paper he intended to submit for publication in the *Geographical Bulletin* was never completed, due to the pressure of other responsibilities.

9   The Geographical Branch Baffin Island project confirmed this hypothesis. Subsequently, deep drilling on the Penny and Devon Island ice caps arrived at similar conclusions, based on dating bottom layers of both.

10 Dr. W. van Steenburgh was director general of scientific services (later, deputy minister). As such, he was the ultimate authority. His scientific training as a biologist was especially important from my point of view as he had no direct scientific affiliation with any of the five branches of the department.

2 | Reconnaissance 1961: Learning about Airborne Support

1   Hugh had been a member of the 1953 AINA expedition to the Penny Ice Cap and had completed his doctorate at McGill University (1954) on the geomorphology and glacial history of Pangnirtung Pass.

2   Captain R. M. Southern, after a long career with the Royal Navy during which he specialized in hydrographic survey, was recruited by the Arctic Institute of North America as principal investigator to manage a contract with the Canadian Hydrographic Service. This was to produce *The Arctic Pilot and Sailing Directions*. Capt. Southern was a taskmaster of the first order, but a very fine character, and I was proud to work under his strict discipline.

3   The lab's name (McGill Sub-Arctic Research Station) and functions changed in 1971.

4   This level of precision was not available until the following year, when our Surveys and Mapping Branch had begun work on 1:50,000 scale topographic maps.

5   I justified this name proposal to the Canadian Permanent Committee on Geographical Names in the following manner. All the time we spent in the upper watershed of the Isortoq and Rimrock rivers, we had to contend with wretched drinking water—most surfaces had a layer of fine silt. Rimrock Lake itself was clouded with silt, the result of being in the basin of a series of former ice-dammed lakes. Once across the local divide, we had beautiful clear water—hence the "Freshney River." So I was able to conform to the formal rules of nomenclature that required "local" justification. It was not entirely incidental, however, that the beautiful trout stream where I used to fish as a teenager and which flowed into Grimsby's River Head in North Lincolnshire was called the Freshney.

6   We had pondered over a suitable name for this strategically located lake that would prove vital for future operations. It was dammed up against the ice cap. The several higher abandoned shorelines ringing the lake indicated recent loss of level, and a series of glacial drainage channels down the side of and beneath the Lewis Glacier pointed to the path of the escaping water. We were stumped for a really good name until Pauline made a play on its uncertain future—while also provoking a degree of gallows humour—and proposed "Flitaway." In 1963, the lake gave us a scare with an abrupt fall in level, but it was still in existence in 2013, although considerably smaller.

7   Landing a floatplane required a much shorter distance than takeoff, as the pressure of the water on the pontoons while landing rapidly checked the forward motion, serving as a brake.

8   It was at this camp that we had to arrange for Peter to transfer to Vic's operation because of continued aircraft problems and shortage of food—hence "Separation Lake."

9   Now, a half-century later when climate warming is a major newsmaker, there is a precise basis for determining the absolute amount of ice-margin retreat.

10 We surveyed from the highest indications of contemporary tide level to the highest signs of wave action on the glacial moraine (i.e., the "marine limit").

11 John decided that the tempting Arctic hares should be regarded as a source of food; we were not quite at the starvation point, but close enough. He took out the rifle and proceeded as the "great white hunter." Lying full length and carefully sighting, he squeezed the trigger. His target simply hopped a few feet away and again remained still. Another shot, and the same result, though John swore that he could not have missed. On his third try, the hare collapsed. On retrieving it, however, John found that that he had scored three direct hits on the poor thing: each shot had passed right through the hare.

12 Our field rations were very spartan, designed for cooking on one-pint primus stoves: pemmican, relieved on Sundays by corned beef and (infrequent) cans of fruit in syrup, porridge oats, lots of chocolate, canned sardines, dried fruit, canned Maple Leaf butter, sugar, dried soups, ships biscuits, tea, Nescafé, cans of condensed milk, powdered milk (Klim), and one bottle of Hennessey cognac. The much superior freeze-dried food was just coming onto the market; it was in the face of threat to revolt by the student assistants in later years that I converted to freeze-dried food.

13 Before our departure, Station Chief Lou took me aside: "Advice for next year, old chap! Please inform by radio whoever is station chief when you plan a flight to any of the DEW Line stations. Then we can record your presence. This damned radar isn't good enough to spot you. . . . And don't tell the Russians!"

3 | Building the Team and Developing Credibility

1   In part, this demonstation of progress in applied economic geography was considered essential to our claim that a strong Geographical Branch could undertake research of immediate practical or economic value to Canada. For example, an early undertaking was a study of the dates of freeze-up and ice breakup of the St. Lawrence Seaway; as a result, the seaway authority was able to extend the shipping season by several days, a decision worth many millions of dollars.

2   The situation was further exacerbated by the attitudes of several of the senior members of head office. To provide one disturbing example, I was asked by the director of personnel how I could expect exceptions regarding specializations when, in his recollection, competence in geography simply entailed memorizing the names of capital cities, major rivers, and items of trade between different countries!

3   The Abisko Symposium, led by Professor Gunnar Hoppe, involved a week of field excursions based on Abisko, in Arctic Sweden. It was part of the 1960 International Geographical Congress, with Stockholm as the primary meeting place. The symposium, which attracted most of the world's leading glacial geomorphologists, involved field demonstration of the great advances in Scandinavian research—much of which overturned the standard thinking of the time. The outstanding weather and exceptional conviviality of the symposium led to many lifelong international friendships and collaborations.

4   Professor Hoppe had invited me to dinner prior to my lecture. Inevitably, we discussed the nunatak hypothesis and agreed to disagree. In his summing up after my lecture, he stated that, as he was certain the entire audience would oppose my comments about the nunatak hypothesis, no questions on that topic would be allowed. He was more concerned about my description of large, open, ice-dammed lakes in northeastern Labrador-Ungava, especially since earlier work in Scandinavia had presented similar conclusions but those had been recently overturned. There followed a vigorous question-and-answer session; I persuaded most in the audience that my Labrador-Ungava evidence needed careful consideration, at least. In 2002, I was delighted to receive a copy of a Stockholm doctoral dissertation from its author, Krister N. Jansson, together with greetings from the renowned Professor Hoppe. The frontispiece was devoted to a long quotation from the paper that had been the basis for my lecture in Stockholm in 1961: "The damming by ice of lakes several thousands of square miles in extent and up to 1,000 feet deep, would have required a massive ice barrier to shut off their natural drainage outlet—Ungava Bay. Allied with the conclusions drawn from the Torngat Mountains, this results in the postulation of a major cupola of the inland ice over the present site of Ungava Bay until relatively late in glacial time. This concept is somewhat alien to current hypothetical thought (Ives, 1960[b])."

5   Gunnar had invited me home to dinner to meet his wife, Britta, and their three children. Britta was delightful although more than a little dubious of the sudden news that she was about to emigrate to Canada. Gunnar extolled the virtues of their future prospects: there was such a lot of ice on Baffin Island—much more than Norway and Sweden combined! After dinner, and somewhat whimsically, Britta invited me to see her refrigerator. This surprised me. But when she opened the freezing unit, I could not restrain a laugh. "Look!" she said, "All the space is taken by Gunnar's ice samples. You will have to watch him very carefully or else he will try to bring half the Barnes Ice Cap down to Ottawa." Little did I know then how prescient she was (see chapter 4).

6   The actual origins and vicissitudes of glaciological research within the Canadian federal government are too complicated to relate within the context of this book and go well beyond its objectives; nevertheless, an outline is warranted. Prior to my appointment with the Geographical Branch, there had been a considerable glaciological research effort, some of which was ongoing. As this was related to the emerging Cold War, it focussed on the High Arctic and the Arctic Ocean—ice islands, the Ellesmere ice shelf, and Operation Hazen—and was primarily located within the Defence Research Board, under the leadership of Geoffrey Hattersley-Smith and Moira Dunbar. In 1960, with Fred Roots as director, the Polar Continental Shelf Project (PCSP) being established within the Department of Mines and Technical Surveys was also aimed at the High Arctic, but as a combined logistical and research operation. Several new research positions had been created, although they were placed administratively within different branches of the Department of Mines and Technical Surveys; two were allocated to the Geographical Branch. One was used to recruit Keith Arnold, who began glaciological research on Meighan Island. He became a member of the branch's Glaciology Section upon its establishment. The other was used by Geographical Branch director Norman Nicholson to recruit a human geographer who never even went North. Regardless, the Glaciology Section was the first formally established unit for glaciology within the federal government. Later, in the 1970s, the research wing of the PCSP expanded and Roy Koerner began a long and illustrious career working on the Devon Island ice cap and, subsequently, in conjunction with several others, throughout Canada. The original Glaciology Section has survived a tangle of bureaucratic manipulations. It is currently, perhaps ironically, part of the GSC, where its existence is once more precarious in the wake of the federal government's efforts to reduce expenditures.

7   Mary decided to support my insertion of the Scandinavian language requirement into the formal job description, given the need for a Scandinavian language related to the greatly advanced research that had been completed in Norway and Sweden, much of it published in the home languages. Also, the pre-eminent research journal was *Geografiska Annaler*, edited and produced in Stockholm. It was a surprise to me that this critical journal was not available in the otherwise impressive library of the GSC.

8   While I anxiously restrained the mountaineering instincts of many members of the summer field teams, Uwe Embacher (following a discrete lapse of time) did make the first ascent of the highest of the Inugsuin Pinnacles in 1977—solo! (see Fig. 31)

9   Two weeks before writing this passage, I attended a lecture by Professor Michael Church at Carleton University. He had just retired from a long and distinguished career on the faculty at UBC (his doctorate had been based on fieldwork in Baffin Island). My colleague Professor Chris Burn introduced Mike prior to the formal presentation and mentioned some of his many distinctions, including his recent appointment as senior editor of the *Journal of Geophysical Research*. But that was only the beginning. The following week, Mike arranged for me to be invited to the ceremony in Rideau Hall at which the Governor General presented him with the Massey Medal, one of the most distinguished awards conferred by the Royal Canadian Geographical Society. During the reception that followed, I was delighted to be introduced to Mike's charming mother, who kindly voiced the pleasure of meeting, after so many years, the fellow who had enticed her boy to Baffin Island.

10   While vertical air photography of Canada had begun in the early part of the twentieth century, given the immense area involved, progress was extremely slow. Towards the end of the Second World War, tri-camera (trimetragon) operations were introduced, using three interconnected cameras, the central one set to film vertically and those on either side taking oblique photographs. While this method sacrificed accuracy, progress was accelerated by six times. Flying height was set at ten thousand feet asl, and in rugged terrain such as Baffin Island, this restricted the mapping scale to 1:500,000 (eight miles to the inch). The RCAF was responsible for the photography while the Department of Mines and Technical Survey stored the negatives and produced the maps. The photographic operations involved 550 personnel and thirty-three aircraft in the field and several hundred ground personnel stationed at Rockcliffe. Three squadrons were involved, with 408 Squadron, flying eight long-range Lancaster X aircraft, undertaking the tri-camera operations. Baffin Island was photographed completely in 1948. A full account is available at http://pubs.aina.ucalgary.ca/arctic/Arctic3-3-150.pdf.

11   The entire field area was eventually covered by topographic maps at a scale of 1:50 000, unprecedented for an uninhabited and remote Arctic region in that period, although the maps were not printed and not available for our actual fieldwork.

4 | Baffin 1962: Ice Mining on the Barnes Ice Cap

1   Hans (Housi) Weber was the only available glaciologist with gravimetric and seismological training. He had been a member of the 1953 AINA expedition to the Penny Ice Cap and was also a personal friend who had been keen to join us. Our wives, Meg and Pauline, swapped baby-sitting duties during our absence. Later, Housi made the first gravity measurements at the North Pole, and his son Richard undertook record-making over-ice expeditions to the Pole and across the entire Arctic Basin.

2   George recounted many years later that the surprise visit by Murray Watts was related to the prospector's discovery of the Mary River iron deposits of northern Baffin. Murray's Cessna pilot, Ron Sheardown, had amazed George and Mike by making their initial landing on a very shallow stretch of the upper Ravn River (then called Pilik River) where George and Mike were camping. Murray sheepishly confessed later that he had "liberated" one of George's air photographs that they had found in his vacated base camp at the east end of Pilik Lake (now officially, Angaljurjualuk Lake). The air photograph was used to plot what he called "interesting showings" on Nuluujaak Mountain. To obtain the photograph, he must have opened unlabelled and tightly strapped fibre cases. Murray never said a word about iron ore, although it became widely known subsequently that he had discovered one of the world's highest-grade deposits of magnetite. However, after what seems to have been an interminable delay (because of inaccessibility of the High Arctic and current easier ice conditions due to climate warming; see the Baffinland Iron Mines website for fascinating details), the first shipment of ore was finally transported in 2008 via Milne Inlet to a Thyssen blast furness in Germany. It attracted keen attention because of its unusually high grade, although the economic downturn that marked 2008 resulted in a major hiatus.

3   In practice, while in the hands of several different organizations, the research is still ongoing.

4   Gunnar's attempt to move a large chunk of the Barnes Ice Cap to Ottawa for detailed microscopic examination and radiocarbon dating turned out to be a very mixed "success." The positive aspect was that examination of the crystal structure present in the ice core of the moraine proved emphatically that the original source was not glacier ice. The very small crystals—in contrast to the much larger glacier ice crystals—proved, as in Swedish

Lapland, that the original source was snow that had accumulated along the margin of the ice cap and been buried by the buildup of the end moraine and subsequently compressed. The early radiocarbon dates were illogically "old" when compared with other field investigations away from the ice cap. After much investigation, it had to be assumed that the anonymously "old" dates must have resulted from minute shreds of plastic scraped by loose pieces of ice from the container bottles while the samples were being backpacked from the "mining" site to the waiting Beaver aircraft. This plastic material (actually, "dead" carbon), if radiocarbon dated alone, would have yielded an infinite age. The mix of even minute particles of the plastic—with a very small amount of naturally occurring carbonaceous material—within the ice that had been collected would yield an abnormally "old" age.

5   Gunnar also managed a second "ice cream feat." The geography department at Stockholm University had developed a tradition of serving ice cream with lunch on Fridays. This became a playful contest—of who could manage to provide ice cream originating the greatest distance from the department, e.g., from the Stockholm suburbs, Uppsala, Göteborg, Oslo—and the competition was intense. Gunnar managed to win the contest by delivery of ice cream from Baffin Island, courtesy of the chief cook at Fox-2 and the indulgence of flight attendants of three commercial airlines who succumbed to Gunnar's persuasiveness. However, that caused total collapse of the contest! Who could possibly compete with Gunnar?

5 | Expanding Baffin Research and Wider Reconnaissance, 1963

1   The core of Roland Beschel's doctoral dissertation was the study of the effects of air pollution on lichens in several Austrian cities. His research included measurement of the diameters of several species of rock lichens on tombstones in churchyards high in the Tyrolean Alps. The tombstones, of course, had the dates when they were erected carved into them, so he was able to plot the diameter of lichen species against their age, based on the dates when the stones had been erected, giving a timescale of nearly a thousand years. The result was a lichen growth curve, the longest section of which was a straight line. He next measured the diameters of the same species of lichens that were growing on boulders along the surface of moraine ridges fronting nearby glaciers. Through comparison with his tombstone plots, he obtained a good approximation of the actual dates of local glacier fluctuations. The brilliance of his method—which he called lichenometry—rested on the premise that it would be the single largest lichen thallus on some geomorphic or anthropologic feature that would give the best indication of age. He

later applied this technique in western Greenland and on Axel Heiberg Island. His method of maximum "diameters" has stood the test of thousands of applications. While this is an extremely simplified account, it may serve as a general description of his method. It was a very important innovation for the time, as the near absence of carbonaceous material in the Arctic (inland from the coasts, where seashells and driftwood on raised beaches provided carbonaceous material) severely limited the means of dating. Roland's premature death at the age of forty-two was not only a great loss for all of us pursuing Arctic natural science but also the tragic departure of a most engaging personality. Over the last twenty years or so, much more intensive work has been completed on the theory of lichenometry and, as so often happens in such cases, there are several complications, although the basic tenets of Beschel's original work stand.

2   The light-toned areas have come to be referred to as "lichen-free" areas. This became a convenient form of expression that later (post-1967) caused a degree of confusion. The light-toned areas were not actually lichen-free; they were light in tone because of small-diameter lichens and very limited total lichen cover.

3   This involvement with Pat Webber led to a remarkable series of events that culminated in his succeeding me as director of the University of Colorado Institute of Arctic and Alpine Research in 1979.

4   The first sample of Barnes Ice Cap end moraine ice had proved to contain insufficient carbonaceous material to permit effective C_{14} dating, so a much larger sample was collected during the 1963 field season.

5   The term "black frost" is used by British fishing fleets in Arctic waters. It describes a condition that occurs during winter when trawlers are fishing in high winds and low temperatures. The heavy spray freezes on the windward superstructure and, in severe conditions, causes the vessel to capsize.

6 | Initiatives and Growth in Baffin Operations, 1963–1964

1   In practice, I became deputy chair, in deference to Mr. J.-P. Drolet, who had recently joined the department as assistant deputy minister.

2   The Geographical Branch provided, from its central budget, Canada's annual contribution to the IGU. This also entailed responsibility for the national committee, composed primarily of university geographers; facilitation of committee meetings; and a lead role in determining allocation of funds to help Canadian geographers attend the International Geographical Congress held every four years.

3   At the time this manuscript was undergoing final corrections, this most promising encouragement accidentally came to be discussed with an old departmental colleague, who assured me that the Pleistocene geologists of the GSC were most certainly not enthusiastic about Dr. Harrison's proposal. Therefore, I have to assume that Dr. Harrison and Dr. van Steenburgh had been confident that they could have finessed the transition. We live and learn!

4   For instance, as chair of the United Nations University Advisory Committee on Research he strongly influenced my appointment in 1978 as coordinator of its "mountain project," allowing me the very real pleasure of working under his general supervision in one of the greatest adventures of my career. See *Sustainable Mountain Development: Getting the Facts Right* (Ives, 2013).

5   Dr. Cuchlaine King was my undergraduate tutor at the University of Nottingham. She became a key member of the two university student expeditions to southeast Iceland of which I was leader. We published four joint papers in the *Journal of Glaciology* and she became a close friend. It was Cuchlaine and her Cambridge women undergrad assistants, as well as my wife, Pauline, who together convinced me that the criticism of women in harsh field conditions was prejudicial nonsense.

6   Gerasimov actually generated a turmoil at the Congress by publicly condemning Britain for "deliberately" withholding visas from the East German delegation; he followed up by resigning from his president-elect position, thereby ensuring that the next Congress was held in New Delhi instead of Moscow. He did host the 1976 Congress in Moscow and invited me as a special guest. Also, following a visit to our home in Boulder, Colorado, Gerasimov pressed me to move with my entire family to Moscow and enjoy a much higher standard of living, courtesy of the Soviet regime. Many years later, I was invited to Moscow to read a eulogy in memory of his life's work.

7   Experience on the DEW Line saw impeccable behaviour by the "deprived" male staff. The women's physical stamina fell within the median of that of the men, although few could out-walk Cuchlaine. In my estimation the mixed party, as in Iceland, ensured enhanced conviviality and contributed to the seriousness of the students' approach to their work. The DEW Line senior staff fell over themselves to be helpful, and I found that the women, in this context, were worth their weight in gold. Ed Smith and Wendy Jocelyn, and Mike Church and June Ryder, were Baffin party members who subsequently married.

8   In effect, McGill University and AINA were the primary institutions involved, although the fact that Joan was on the payroll of the Geographical Branch ensured its inevitable involvement.

9   Royalty were exceptions; the closest of these to our Baffin Island field sites was Prince Charles Island. And there was one further, very unusual exception. On the occasion when our friend and colleague Professor J. Ross Mackay was awarded the Massey Medal, there was a reception at Rideau Hall hosted by the Governor General and Mrs. Vanier. (I had been invited because of my recent appointment as assistant director of the Geographical Branch, where Ross was spending his sabbatical.) Over kindly chit-chat with my wife, Pauline, Mrs. Vanier said that surely I would now be able to name a mountain or an island in the Arctic for her. Pauline explained the strict rules of nomenclature. Mrs. Vanier won our hearts by pointing out that there was a small island in the High Arctic named for her—and as Pauline Vanier, she could share it with Pauline Ives.

10 George Falconer was my indispensable assistant in this process. And we were not completely devoid of a sense of mischief in selecting the almost sacred symbol of Samuel de Champlain, appointed Royal Geographer to Canada by King Henry of Navarre in 1603. In part, we were reacting to being told by our senior and sometimes overbearing neighbours at 601 Booth Street (GSC)—that their truly ancient lineage extended back to before Confederation (only just).

11 Very long and obscure names were not unusual during the nineteenth century, as the Royal Navy attempted to find the Northwest Passage (e.g., place names such as "Collingworth, His Farthest"). The "Sons of the Clergy," however, was a sarcastic jibe aimed at the Scottish Episcopal Church clergy, who were known, apparently, for their very large families. The current name is shorter: Sons of the Clergy Islands, Nunavut.

7 | Glaciology in the Rockies Added to Baffin Studies, 1965

1   This gave us two party members with identical names, so I will refer to our pilot as David and to his geographer namesake as Dave.

2   In those relatively early days of radiocarbon dating (C_{14}), 40,000 to 50,000 years BP (before present, meaning, before 1950) was at the limit of the method; therefore, barring any contamination from very "old" carbon, such as coal in an extreme case, or minute pieces of plastic in the case of Gunnar's ice-cored moraines, it was normal to take the result as implying "older than 50,000" years before 1950.

3   Many of the photographs were used in two filmstrips, commissioned by the National Film Board (NFB), that were designed for teaching physical geography in high schools: *Glaciers* and *Glacial Landforms*. The NFB

supplied photographic film and undertook carefully controlled laboratory processing.

4   In fact, George and I made a special day trip to Toronto to visit Murray Watts and staff of the newly formed Baffinland Iron Mines Company. We were seeking collaboration in the north of Baffin Island. The trip also allowed us to visit a large photographic store on Yonge Street and inspect a Hasselblad system, there being none available in Ottawa at that time.

5   While it had appeared to me at the time that we were above a nearly complete cloud cover, David (personal communication, November 7, 2010) subsequently explained that he always had some ground contact (visual flight rules [VFR]) and never would have operated without. In fact, in Baffin he often had to use the opposite strategy: to hug the ground at only a couple of hundred metres while staying beneath a cloud cover or fog bank. "I think particularly of the low-level approach into Inugsuin down the steep ravine valley from the west that we had to use many times to 'get home to dinner,'" he recalled. "I would practice flying those low-level routes on my own on bright sunny days, aware that at some time in the near future, I'd have to know every nook and cranny to do it under darker clouds in (relative) safety. Of course, also, at such low altitudes, autorotation would have been highly tricky!"

6   The National Advisory Committee membership included: Dr. L. Beauregard, chair, Inst. de Géographie, Université de Montréal; Dr. J. W. Birch, chair, geography dept., University of Toronto; Dr. E. Christiansen, Saskatchewan Research Council; Dr. J. G. Fyles, Pleistocene Section, GSC; Dr. F. Grenier, chair, Inst. de Géographie, Université Laval; Dr. J. D. Ives, director, Geographical Branch, *ex officio*; Dr. G. Jacobsen, Tower Construction Company, Montreal; Dr. T. Lloyd, chair, geography dept., McGill University (Montreal); Dr. J. R. Mackay, geography dept., University of British Columbia; Mr. P. Marchant, British Newfoundland Corporation; Mr. M. K. Thomas, Meteorological Branch, Dept. of Transport; Dr. K. W. Walter, geographic advisor, Imperial Oil; Dr. T. Weir, chair, geography dept., University of Manitoba; Dr. W. C. Wonders, chair, geography dept., University of Alberta (Edmonton). Secretary *ex-officio*: Mrs. Alexandra Cowie, chief administrative officer, Geographical Branch.

7   Alex was also a great asset during branch budget negotiations with head office finance personnel—competent staff of the old school who were totally unaccustomed to arguing with a woman.

8 | Summit Experiences and East Coast Research, 1966

1   In David's words, "One had to try to 'see' and 'feel' the updrafts and downdrafts and have total respect for the power of nature that 'lifts you up' or would dash you to pieces against the rocks if you became over-confident. This is no country for show-offs!" (D. Harrison, personal communication, 7 November 2010).

2   David later said of the ice axe, "I am glad I never had to use it—I wouldn't have had a clue. I suspect you only had one set anyway, knowing that I would probably elect to stay on top with the helicopter [rather] than to keep you company on the climb down" (Ibid.).

3   I recently found this photo on the web, used without permission and with no caption. Since I had not had the original transparency scanned, it is a mystery how it escaped my copyright coverage. However, I am happy to accept its improper use as an acknowledgement of my ancient (1966) photographic judgement.

4   A further comment from David: "I can visualize the adrenalin-fuelled excitement and tension in the cabin as I write this. Two mature explorer-adventurers weighing adventure and safety with the pilot having the last word (but I was *so* aching to land there!)." (7 November 2010).

5   The problem of interpreting the small summit blocks that are probably glacial erratics is rendered the more complex for the botanist who is intent on "proving" that vascular plants survived throughout the ice ages on mountaintops that were not submerged by the continental ice sheets. Many of the higher mountaintops in Baffin Island are topped by thin ice carapaces, so thin that they are almost certainly frozen to the underlying surface and, if melted away, would likely leave no trace of their former presence. Even if no presumed glacial erratics are found, nor other indications of glacial erosion, the ice carapaces, so common under today's climate, surely existed throughout the ice ages. In other words, my findings in Baffin Island in the 1960s, along with comparable earlier observations from the Torngat Mountains (Ives, 1958, 1960b), gave some indication of the probable highest former ice sheet surface (i.e., close to and/or slightly above the high summits). Survival of vascular plants, or any life form, would be unlikely unless seeds can remain viable for many thousands of years. The probability of such plant viability—at least for hundreds, if not thousands, of years—was an issue already raised by Falconer (1966) regarding Baffin Island, and by others looking elsewhere (but see chapter 11).

6   On reading an early draft of this book, David Harrison made the following comment: "Very good reaction, Jack—I was proud of you—a hole in the roof (which would probably have exploded the whole bubble) would not have done either of our careers much good. In fact,

this entire story would have been decidedly posthumous" (7 November 2010). I took this to be another relic of David's Brit sense of humour.

7 I proposed the name "Shadow Mountain" with the explanation that I had had difficulty obtaining a photograph of its impressive north face, as it always was in shadow when I was in a position to photograph. I also had an ulterior motive that would have broken the federal rules of nomenclature: the skipper of the Grimsby fishing trawler who had introduced me to the Arctic at the age of fifteen (Arctic Norway and Svalbard) had the nickname "Shad," or "Shadow," from being so skinny as a trawler mate prior to the Second World War. The *Grimsby Telegraph* wrote a good headline for the article I submitted in tribute to the man who had indelibly influenced the course of my career: words to the effect of "Baffin Island Mountain Named 'Shadow Mountain' in Honour of Skipper Arthur Phillipson, a Rock of a Man."

9 | Last Year of Baffin Island Activities by the Geographical Branch, 1966–1967

1 Publication of a desk-size atlas was Gerry Fremlin's idea and very much in keeping with my sense of urgency in making the Geographical Branch significant to Canada at large. The previous *National Atlas of Canada* had involved a large binder from which individual maps could be inserted and removed. This facilitated the replacement of outdated maps with new editions, or simply the addition of new maps; however, it was not easy to carry around. The new desk atlas was to be bound with a hard cover and of a size that students could carry to school and others could keep in their homes like any other atlas. In this way, we hoped that the new atlas—which covered the geography, natural resources, economy, and land use of Canada—would become the best possible tool for alerting the public both to the fascinating details of our immense country and to the existence of a branch of government that was doing something of great educational value. However, it was by no means considered a replacement for another edition of the main *Atlas of Canada*. Nevertheless, there are intractable problems associated with producing a national atlas for a country the size of Canada, and several changes occurred during the following decades. The current approach is digital: see http://www.nrcan.gc.ca/earth-sciences/geography/atlas-canada.

2 This was a time of growing, but often begrudging, attempts by the federal government to reach out to our French Canadian partners. I thought the formal acknowledgement of Samuel de Champlain as the founder of Canada made eminently good political sense as well as being a most appropriate recognition, yet my proposal that a postage stamp honouring Champlain be issued

during the 1967 run-up to Canada's centennial was ignored.

3 Members of the Canadian National Committee of the International Geographical Union had been urging me for several years to seek departmental support (which meant a considerable level of funding) so that a formal invitation could be extended during the next Congress (to be held in 1968 in New Delhi). This had put me in additional conflict with Jim Harrison and the GSC. Jim insisted that the timing conflicted with his own determination to seek departmental support for a similar invitation from Canada to the International Union of Geological Sciences. There would not be room in the departmental budget for two international meetings the same year.

4 The original quotation from Sheridan used "geometry," not "geography," although this poetic license had often been taken by geographers.

5 By the time of Mike's evacuation, however, I was well into my new task of reformulating the Institute of Arctic and Alpine Research, to become known internationally as INSTAAR, in Boulder, Colorado.

6 John's reconnaissance work on Broughton Island provided the foundation for the first Baffin Island expeditions organized by INSTAAR from Colorado.

7 Roger Barry joined me in Boulder, Colorado, the following year.

8 Pat Webber succeeded me in 1979 as director of INSTAAR.

9 The word *ookpik* means "snowy owl" in Inuktituk; it also refers to one of the beautiful Inuit handicrafts of the time—a small handmade snowy owl doll fashioned from sealskin.

10 In 1975, when I applied for a Guggenheim Fellowship, Jim Harrison enthusiastically accepted the task of being my principal referee. The following year, as assistant director general of UNESCO, he proved to be a prime influence on my appointment as coordinator of the United Nations University mountain program. As he was chair of the UNU advisory committee, during official visits to headquarters in Tokyo he joked that, finally, he had made sure that I was working for him (see Ives, 2013).

11 | Assessment of the Scientific Results

1 *Nunatak* is an Inuit/Greenlandic word meaning "a mountain surrounded by ice." The nunatak hypothesis originated in the late nineteenth century, when intensive botanical work in Scandinavia led to the realization that the peculiar distribution of a large group of arctic-alpine plant species appeared to require that they had survived the ice ages on ice-free areas in the coastal mountains. This concept became the source of extensive controversy

that has not been resolved even more than a hundred years later. The hypothesis was hotly contested by most geologists and geographers in Norway and Sweden; the dispute was extended to North America and figured extensively in our Baffin Island research.

2 This is a rather complicated topic. For instance, during the glacial maxima the enormous masses of ice that occupied vast areas of the earth's surface caused significant depressions in the earth's underlying crust as well as lowering the sea level. As the ice sheets melted during the closing phases of each ice age, world sea level (or, eustatic sea level) rose. The different timing between eustatic sea level rise and differential isostatic rebound of the formerly depressed sections of the earth's crust produced the complex relationships that became worldwide objects of scientific research, including the Baffin Island work discussed here. It should be pointed out that areas formerly mantled by thick ice-age ice are still slowly rising. The coastal areas of Hudson Bay, Foxe Basin, and the Gulf of Bothnia are prime examples. See also Savelle and Dyke (2014) for their use of dated former sea levels around Foxe Basin and other areas as a means of determining elements of Arctic archaeology.

3 The recognition of former high sea levels associated with the thinning and retreat of the last great ice sheets resulted from early observations on raised marine beaches in the Baltic Sea, especially around the Gulf of Bothnia. The maximum height of the former sea level stands was seen to increase from southern Sweden and Finland northwards. A distinction was quickly made between the multiple flights of intermittent beach terraces, strandlines (actual continuous former sea levels), and the upper marine limit (the highest point reached by salt water at any location, not necessarily contemporaneously). Precise instrumental survey of the strandlines demonstrated that the isostatic uplift of the land was greatest around the north end of the Gulf of Bothnia. In other words, the strandlines were tilted up in the direction of the former greatest thickness of the Fenno-Scandinavian Ice Sheet. The implications of these early discoveries were quickly transferred to North America. Identification of actual strandlines and their tilt, however, was delayed due to the lack of topographical maps, although tilted and raised lake shorelines of former freshwater lakes dammed against the southern margins of the Laurentide Ice Sheet were mapped as early as the 1930s and 1940s. Løken's work (1960, 1962) in the Torngat Mountains marked the first occasion when strandlines and their tilt

were identified in northern Canada. Ives (1958) began a similar survey, but of freshwater lake shorelines, in northeastern Labrador-Ungava—work that was continued by Matthew (1961), Harrison (1963), and Barnett and Peterson (1964), all as part of the McGill Sub-Arctic Research Lab program.

4 Wordie (1938) had reported distinct marine features up to about sixty metres above present sea level in the outer fiords of the Baffin Island northeast coast. Much higher terraces, seen from shipboard but not inspected directly, were noted to have probably been formed as lateral features by former glaciers.

5 As an indication of the level of uncertainty, the then current edition of the *Encyclopaedia Britannica* (1957), in the section on Labrador (which I had been asked to update) contained the claim that the Torngat Mountains rose abruptly from the Atlantic to between 2,000 and 3,000 metres; their actual heights range from 1,400 to 1,600 metres (maximum 1,652 metres).

6 This section has been rather lengthy because I was personally intrigued by the long-continuing controversy and it had engaged my main research energies at the McGill Sub-Arctic Research Laboratory.

7 Personal information from George Falconer, who was directly involved as one of the key scholars to have worked under Tuzo Wilson's leadership.

8 See chapter 3, note 6.

9 In May 2014, I received an email from Mike Demuth explaining that he was just about to leave home to undertake the annual mass balance determination for Peyto Glacier.

10 Gifford Miller and his coworkers collected similar samples of plants (mainly mosses) from beneath the receding margins of thin ice patches over a great expanse of Baffin Island: from north of the Rowley River, in the north, southeastward to the high plateaus between the fiords of Cumberland Peninsula that enter Davis Strait. The 120,000 BP dates from some of these collections have been used to postulate that Baffin Island summers have been warmer in recent years that at any time since the last (Sangamon) interglacial. Dyke (personal communication, June 23, 2014) believes that this inference—that is, that the highest summer temperatures of this 120,000-year period have occurred only during the last decades—has not been confirmed.

REFERENCES

Adams, P. (2007). *Trent, McGill and the North: A story of Canada's growth as a sovereign polar nation.* Peterborough, Ont.: Cover to Cover.

Andrews, J. T. (1963). Cross-valley moraines of the Rimrock and Isortoq river valleys, Baffin Island, N.W.T.: A descriptive analysis. *Geogr. Bull.*, 19: 49–77.

Andrews, J. T. (1966). Pattern of coastal uplift and deglacierization, West Baffin Island, N.W.T. *Geographical Bull.*, 8(2): 174–193.

Andrews, J. T. (1970). A geomorphological study of post-glacial uplift with particular reference to Arctic Canada. *Inst. Brit. Geogr. Spec. Publ., 2*, 1-156.

Andrews, J. T., & Barnett, D. M. (1979). Holocene (Neoglacial) moraines and proglacial lake chronology, Barnes Ice Cap, Canada. *Boreas*, 8: 342–358.

Andrews, J. T., & Falconer, G. (1969). Late glacial and post-glacial history and emergence of the Ottawa Islands, Hudson Bay, N.W.T.: Evidence on the deglaciation of Hudson Bay. *Can. Journ. Earth Sci.*, 6: 1263–1276.

Andrews, J. T., & Smithson, B. B. (1966). Till fabrics of the cross-valley moraines of north-central Baffin Island, N.W.T., Canada. *Geol. Soc. Am. Bull.*, 77: 271–290.

Andrews, J. T., & Webber, P. J. (1964). Lichenometrical study on the northwestern margins of the Barnes Ice Cap: A geomorphological technique. *Geographical Bull.*, 22: 80–104.

Andrews, J. T., & Webber, P. J. (1969). Lichenometry to evaluate changes in glacial mass budgets: As illustrated from north-central Baffin Island, N.W.T. *Arctic and Alpine Research*, 1(3): 181–194.

Baird, P. D., & Ward, W. H. (1952). The glaciological studies of the Baffin Island Expedition, 1950: Part I: Method of nourishment of the Barnes Ice Cap; Part II: The physics of deglaciation of Central Baffin Island. *Journ. Glaciol.*, 4: 2–22.

Barber, D. C., Dyke, A., Hillaire-Marcel, C., Jennings, A. E., Andrews, J. T., Kerwin, M. W., . . . Gagnon, J-M. (1999). Forcing of the cold event of 8,200 years ago by catastrophic drainage of Laurentide lakes. *Nature*, 400: 340–348.

Barnett, D. M. (1977). *Glacial geomorphology in a sub-polar proglacial lake basin* (Doctoral dissertation). University of Western Ontario, London.

Barnett, D. M., & Peterson, J. A. (1964). The significance of glacial Lake Naskaupi 2 in the deglaciation of Labrador-Ungava. *Can. Geographer*, 8: 173–181.

Bell, R. (1884). *Observations on geology, mineralogy, zoology and botany of the Labrador coast, Hudson's Strait and Bay.* Report on Progress, 1882–1884, Vol. 1, Part D, 5–62. Ottawa: Geological Survey of Canada.

Beschel, R. (1957). Lichenometrie im Gletschervorfeld. *Sonderdruck uas dem Jahrbuch 1957 des Versins zum Schutze der Alpenpflanzen und–Tiere München e. V..* Munich 2, Linprunstrasse 50/iv.

Beschel, R. E. (1961). Dating rock surfaces by lichen growth and its application to glaciology and physiography (Lichenometry). In G. A. Raasch (Ed.), *Geology of the Arctic* (Vol. 2) (pp. 1044–1062). Toronto: University of Toronto Press.

Bhatt, U. S., Walker, D. A., Raynolds, M. K., Comiso, J., Epstein, H. E., Jia, G. S., . . . CAVM Team. (2003). Conservation of Arctic flora and fauna (No. 1) [Circumpolar Arctic vegetation map]. Anchorage, Alaska: US Fish & Wildlife Service.

Blackadar, R. G. (1958). Patterns resulting from glacier movements north of Foxe Basin, N.W.T. *Arctic*, 11(3): 156–165.

Blake, W. Jr. (1963). *Notes on glacial geology, northeastern District of Mackenzie* (Paper No. 63-28). Ottawa: Geological Survey of Canada.

Blake, W. Jr. (1966). *End moraines and deglaciation chronology in northern Canada with special reference to southern Baffin Island* (Paper No. 66-26). Ottawa: Geological Survey of Canada.

Blake, W. Jr. (1970). Studies of glacial histories in Arctic Canada: Pumice, radiocarbon dates, and postglacial uplift in the eastern Queen Elizabeth Islands. *Can. Journ. Earth Sci.*, 7: 634–664.

Blake, W. Jr. (1975). Radiocarbon age determinations and postglacial emergence at Cape Storm, Southern Ellesmere Island, Arctic Canada. *Geogr. Ann.*, 57A: 1–71.

Brassard, G. R., Fife, A. J., & Webber, P. J. (1979). Mosses from Baffin Island, Arctic Canada. *Lundbergia*, 5: 99–104.

Brochmann, C., Gabrielsen, T. M., Nordal, I., Landvik, J. Y., & Elven, R. (2003). Glacial survival or tabula rasa? The history of North Atlantic biota revisited. *Taxon*, 52(3): 417–450.

Brown, R. J. E. (1967). Permafrost in Canada [Map No. 1246A]. Ottawa: Geological Survey of Canada.

Callaghan, T. V., Tweedie, C. E., & Webber, P. J. (2011). Multi-decadal changes in tundra environments and ecosystems: Synthesis of the International Polar Year–Back to the Future Project (IPY-BTF). *Ambio*, 40: 555–557.

Canadian Hydrographic Service. (1959). *Pilot of Arctic Canada*. Ottawa: Queen's Printer.

Church, M. (1970). *Baffin Island sandar: A study of Arctic fluvial processes* (Unpublished doctoral dissertation). University of British Columbia, Vancouver.

Church, M. (1972). *Baffin Island sandurs: A study of Arctic fluvial environments* (Bull. No. 216). Ottawa: Geological Survey of Canada.

Church, M., & Gilbert, R. (1975). Postglacial fluvial and lacustrine environments. In A. V. Jopling & B. C. McDonald (eds.), *Glaciofluvial and Glaciolacustrine Sedimentation* (pp. 21–100). Soc. Economic Paleontologists and Mineralogists, Spec. Publ. No. 23.

Clarke, G. C. K., Leverington, D. W., Teller, J. T., & Dyke, A. S. (2004). Paleohydraulics of the last outburst flood from glacial Lake Agassiz and the 8200 BP cold event. *Quaternary Sci. Rev.*, 23: 389–407.

Coleman, A. P. (1920). Extent and thickness of the Labrador Ice Sheet. *Bull. Geol. Soc. Am.*, 31: 319–328.

Coleman, A. P. (1921). *Northeastern part of Labrador and New Quebec* (Memoir 124). Ottawa: Geological Survey of Canada.

Coleman, A. P. (1926). Pleistocene of Newfoundland. *Journ. Geology*, 34: 193–223.

Craig, B. G., & Fyles, J. G. (1960). *Pleistocene geology of Arctic Canada* (Paper No. 60-10). Ottawa: Geological Survey of Canada.

Dahl, E. (1955). Biogeographical and geological indications of unglaciated areas in Scandinavia during the ice ages. *Bull. Geol. Soc. Am.*, 66(12): 1499–1520.

Dahl, E. (1961). Refugieproblemet og de Kvartaergeologiske metodene. *Svensk Naturv*, 14: 81–96.

Daly, R. A. (1902). The geology of the northeast coast of Labrador. *Bulletin of the Museum of Comparative Zoology*, 38: 205–270.

De Geer, G. (1912). A geochronology of the last 12,000 years. 11th Internatl. Geol. Congress, Stockholm. *Compt Rendu*, 1: 241–258.

Dyke, A. S. (2004). An outline of North American deglaciation with emphasis on central and northern Canada. In J. Ehlers & P. L. Gibbard (Eds.), *Quaternary Glaciations–Extent and Chronology, Part II* (pp. 373–424). Amsterdam: Elsevier, Developments in Quaternary Science 2.

Dyke, A. S., Andrews, J. T., Clark, P. U., England, J. H., Miller, G. H., Shaw, J., & Veillette, J. J. (2002).

The Laurentide and Innuitian ice sheets during the Last Glacial Maximum. *Quaternary Sci. Rev.*, 21(1–3): 9–31.

Falconer, G. (1962). *Inventory of Canadian glaciers, part 1: Northern Baffin and Bylot islands, N.W.T., Canada* (Paper No. 33). Ottawa: Geographical Branch, Dept. of Energy, Mines, & Resources.

Falconer, G. (1966). Preservation of vegetation and patterned ground under a thin icebody in northern Baffin Island, N.W.T. *Geographical Bulletin*, 8(2): 194–200.

Falconer, G., Henoch, W., & Østrem, G. (1966). A glacier map of southern British Columbia and Alberta. *Geogr. Bull.*, 8: 108–112.

Falconer, G., Andrews, J. T., & Ives, J. D. (1965). Late Wisconsin end moraines in northern Canada. *Science*, 147 (3658): 608–610.

Falconer, G., Ives, J. D., Løken, O. H., & Andrews, J. (1965). Major end moraines in eastern and central Arctic Canada. *Geographical Bulletin*, 7: 137–153.

Feyling-Hanssen, R. W. (1967). The Clyde Foreland. In O. H. Løken (Ed.), *Field report: North-central Baffin Island, 1966* (pp. 35–55). Ottawa: Geographical Branch, Dept. of Energy, Mines, & Resources.

Feyling-Hanssen, R. W. (1976). The Clyde Foreland Formation: A micropaleontological study of the Quaternary stratigraphy. *Marine Sediments*, Special Pub. No. 1, Part B, 315–377.

Fernald, M. L. (1925). Persistence of plants in unglaciated areas of boreal America. *American Acad. Arts & Sciences*, Mem. 15(3): 239–242.

Flint, R. F. (1943). Growth of the North American ice sheet during the Wisconsin age. *Geol. Soc. Am. Bull.*, 54: 325–362.

Flint, R. F. (1947). *Glacial geology and the Pleistocene Epoch*. New York: John Wiley & Sons.

Flint, R. F. (1952). The Ice Age in the North American Arctic. *Arctic*, 5: 135–152.

Flint, R. F. (1957). *Glacial and Pleistocene Geology*. London: Chapman & Hall.

Flint, R. F. (1971). *Glacial and Quaternary Geology*. New York: John Wiley & Sons.

Flint, R. F., et al. (1945). Glacial map of North America [Map]. *Geol. Soc. Am.,* Spec. Paper No. 60, Pt. 1.

Frodin, G. (1925). Studien über die Eisscheide in Zentral-skandinavien. *Bull. Geol. Inst.*, 19: 141–258.

Gjessing, J. (1960). *Isavsmeltningstidens Drenering*. Oslo: Ad Novas det Norsk Geografiske Selskab.

Goldthwait, R. P. (1951). Development of end moraines in east-central Baffin Island. *Journ. Geology*, 70: 737–753.

Harrison, D. (1967a). Relief information unreliable. (Part 1.) *Shell Aviation News*, 345: 12–17.

Harrison, D. (1967b). Relief information unreliable. (Part 2.) *Shell Aviation News*, 346: 16–21.

Harrison, D., & Benjamin, A. (1967). Peak performance. *Shell Aviation News*, 353: 22–24.

Harrison, D. A. (1963). The tilt of the abandoned lake shorelines in the Wabush-Shabogamo Lake area, Labrador. *McGill Sub-Arctic Research Papers*, 15: 14–22.

Henoch, W. E. S., & Stanley, A. (1970). Glacier maps of Canada. *Journ. Glaciol.*, 9(55): 49.

Hoppe, G. (1952). Hummocky moraine regions with special reference to the interior of Norbotten. *Geog. Annaler*, 34: 1–72.

Hoppe, G. (1959). Glacial morphology and inland ice recession in Northern Sweden. *Geog. Annaler*, 41: 193–212.

Ives, J. D. (1957). Glaciation of the Torngat Mountains, northern Labrador. *Arctic*, 10(2): 67–87.

Ives, J. D. (1960a). Former ice-dammed lakes and the deglaciation of the middle reaches of the George River, Labrador-Ungava. *Geographical Bulletin*, 14: 44–70.

Ives, J. D. (1960b). The deglaciation of Labrador-Ungava: An outline. *Cahiers de Géographie de Québec*, 4: 323–343.

Ives, J. D. (1962). Indications of recent extensive glacierization in north-central Baffin Island, N.W.T. *Journ. Glaciol.*, 4: 197–205.

Ives, J. D. (1963a). Determination of the marine limit in Eastern Arctic Canada. *Geographical Bulletin*, 19: 117–122.

Ives, J. D. (1963b). Field problems in determining the maximum extent of Pleistocene glaciation along the eastern Canadian seaboard: A geographer's point of view. In A. Löve & D. Löve (Eds.), *North Atlantic biota and their history* (pp. 337–354). New York: Macmillan.

Ives, J. D. (1964). Deglaciation and land emergence in northeastern Foxe Basin, N.W.T. *Geographical Bulletin*, 21: 54–65.

Ives, J. D. (1974). Biological refugia and the nunatak hypothesis. In J. D. Ives & R. G. Barry (Eds.), *Arctic and alpine environments* (pp. 605–636). London & New York: Methuen.

Ives, J. D. (1978). The maximum extent of the Laurentide Ice Sheet along the east coast of North America during the last glaciation. *Arctic,* 31(1): 24–53.

Ives, J. D. (2010). *The land beyond: A memoir.* Fairbanks: University of Alaska Press.

Ives, J. D. (2013). *Sustainable mountain development: Getting the facts right.* Kathmandu, Nepal: Himalayan Assoc. for Advancement of Science.

Ives, J. D., & Andrews, J. T. (1963). Studies on the physical geography of north-central Baffin Island. *Geographical Bulletin*, 19: 5–48.

Ives, J. D., & Harrison, D. (1967). Rotorcraft on research. *Can. Geog. Journ.,* 74(5): 145–151.

Ives, J. D., & Kirby, R. P. (1964). Fluvioglacial erosion near Knob Lake, central Quebec-Labrador, Canada: Discussion. *Bull. Geol. Soc. Am.,* 75: 917–922.

Jacobs, J. J., Herm, R., & Luther, J. E. (1993). Recent changes at the northwest margin of the Barnes Ice Cap, Baffin Island, N.W.T., Canada. *Arctic and Alpine Research,* 25: 241–352.

King, C. A. M., & Ives, J. D. (1956). Glaciological observations on some of the outlet glaciers of southwest Vatnajökull, Iceland, 1954: Part I–Glacier Regime. *Journ. Glaciol.,* 2: 563–569.

Koerner, R. M. (1980). The problem of lichen-free zones in Arctic Canada. *Arctic and Alpine Research,* 12: 87–94.

La Farge, C., Williams, K. H., & England, J. H. (2013). Regeneration of Little Ice Age bryophytes emerging from a polar glacier: Implications of totipotency in extreme environments. *Proceedings of the National Academy of Sciences,* 110(24): 9839–9844.

Lara, M. J. (2012). *Implications of decadal time scale and Arctic plant community change on ecosystem function* (Doctoral dissertation). University of Texas at El Paso.

Lee, H. A. (1959). *Surficial geology of the southern district of Keewatin and the Keewatin ice divide, N.W.T.* (Bull. No. 51). Ottawa: Geological Survey of Canada.

Lee, H. A. (1960, May 27). Late-glacial and postglacial Hudson Bay sea episode. *Science,* 131(3413): 1609–1611.

Locke, W. W. III, Andrews, J. T., & Webber, P. J. (1979). *A manual for lichenometry.* London: Brit. Geomorphological Research Group.

Løken, O. H. (1960). Field work in the Torngat Mountains, northern Labrador. *McGill Sub-Arctic Research Papers,* 9: 61–73.

Løken, O. H. (1962). The late-glacial and postglacial emergence and the deglaciation of northernmost Labrador. *Geographical Bulletin,* 17: 23–56.

Løken, O. H. (1965). Postglacial emergence at the south end of Inugsuin Fiord, Baffin Island, N.W.T. *Geographical Bulletin,* 7(3–4): 243–258.

Løken, O. H. (1966). Baffin Island refugia older than 54,000 years. *Science,* 153: 1378–1380.

Løken, O. H., & Hodgson, D. A. (1971). On the submarine geomorphology along the east coast of Baffin Island. *Can. Journ. Earth Sci.,* 8: 185–195.

Løken, O. H., & Sagar, R. B. (1968). Mass balance observations on the Barnes Ice Cap, Baffin Island, Canada. *Intntl. Assoc. Scientific Hydrology,* Publ. 79 (General Assembly, Bern, 1967–Snow and Ice), 282–291.

Löve, Á., & Löve, D. (Eds.). (1963). *North Atlantic biota and their history.* Oxford: Pergamon.

Löve, Á., & Löve, D. (1974). Origin and evolution of the arctic and alpine flora. In J. D. Ives & R. G. Barry (Eds.), *Arctic and Alpine Environments* (pp. 571–603). London, Methuen.

Low, A. P. (1906). *The cruise of the Neptune.* Ottawa: Government Printing Bureau.

Mannerfelt, C. M:son (1945). Några glacialmorfologiska formelement. *Geog. Annaler,* 27(1–4): 1–239.

Margreth, A., Gosse, J., & Dyke, A. (2014). Constraining the timing of last glacial plucking of tors on Cumberland Peninsula, Baffin Island, Eastern Canadian Arctic. *Geophysical Research Abstracts,* 16, EGU 2014 Preview, EGU General Assembly, 2014.

Matthew, E. M. (1961). Deglaciation of the George River basin, Labrador-Ungava. *McGill Sub-Arctic Research Papers,* 11: 29–45.

Mercer, J. H. (1956). Geomorphology and glacial history of southernmost Baffin Island. *Geol. Soc. Am. Bull.,* 67: 553–570.

Mathiassen, T. (1933). Contributions to the geography of Baffin Land and Melville Peninsula. *Report of the Fifth Thule Expedition, 1921–24* (vol. 1, no. 3). Copenhagen: Gyldendalske Boghandel Nordisk Forlag.

Miller, G. H. (1973a). Variations in lichen growth from direct measurements: Preliminary curves for *Alectoria miniscula* from eastern Baffin Island, N.W.T., Canada. *Arctic and Alpine Research,* 5: 333–339.

Miller, G. H. (1973b). Late Quaternary glacial and climatic history of northern Cumberland Peninsula, Baffin Island, N.W.T., Canada. *Quaternary Res.* 3: 561–583.

Miller, G. H., Wolfe, A. P., Steig, E. J., Sauer, P. E., Kaplan, M. R., & Briner, J. P. (2002). The Goldilocks dilemma: Big ice, little ice, or "just-right" ice in the eastern Canadian Arctic. *Quaternary Sci. Rev.,* 21: 33–48.

Miller, G. H., Lehman, S. J., Refsnider, K. A., Southon, J. R., & Zhang, Y. (2013). Unprecedented recent summer warmth in Arctic Canada. *Geophys. Res. Lett.,* 40: 1–7.

Müller, F. (1962). Zonation in the accumulation area of the glaciers of Axel Heiberg Island, N.W.T., Canada. *Journ. Glaciol.,* 4(33): 302–311.

Odell, N. E. (1933). The mountains of northern Labrador. *Geog. Journ.,* 82: 193–211, 315–326.

Ommanney, C. S. L. (2005). History of glacier investigations in Canada. In R. S. Williams & J. G. Ferridno (Eds.), *Satellite Imagery Atlas of Glaciers of the World.* US Geological Survey Prof. Paper 1386-J-1.

Østrem, G. (1964). Ice-cored moraines in Scandinavia. *Geog. Annaler,* 46: 282–337.

Østrem, G. (1966). The height of the glaciation limit in southern British Columbia and Alberta. *Geog. Annaler,* 48A: 126–138.

Østrem, G., Bridge, C. W., & Rannie, W. F. (1967). Glacio-hydrology, discharge and sediment transfer in the Decade Glacier, Baffin Island, N.W.T. *Geogr. Annaler,* 49A(2–4): 268–282.

Parry, W. E. (1824). *Journal of a second voyage for the discovery of a northwest passage.* London: John Murray.

Paterson, W. S. B. (1969). *The physics of glaciers.* Oxford: Pergamon.

Rowley, G. W. (2007). *Cold comfort: My love affair with the Arctic.* Montreal & Kingston, Ont.: McGill-Queens University Press.

Sagar, R. B. (1966). Glaciological and climatological studies on the Barnes Ice Cap, 1962–64. *Geographical Bulletin,* 8(1): 3–47.

Savelle, J. M., & Dyke, A. S. (2014). Paleoeskimo occupation history of Foxe Basin, Arctic Canada: Implications for the core area model and Dorset origins. *American Antiquity,* 79(2): 249–276.

Schwarzenbach, F. H. (2011). *Botanical observations on the Penny Highlands of Baffin Island.* Norderstedt, Germany: Books on Demand.

Schytt, V. (1949). Refreezing of the melt-water on the surface of glacier ice. *Geog. Annaler,* 31: 222–227.

Sim, V. (1961). Maximum postglacial marine submergence in southern Melville Peninsula, N.W.T. *Arctic,* 14(4): 241–244.

Sim, V. (1964). Terrain analysis of west-central Baffin Island, N.W.T. *Geographical Bulletin,* 21: 66–92.

Sugden, D. E., & Watts, S. H. (1977). Tors, felsenmeer, and glaciation in northern Cumberland Peninsula, Baffin Island. *Can. Journ. Earth Sci.,* 3: 243–263.

Sutherland, P. D. (2009). The question of contact between Dorset Paleo-Eskimos and early Europeans in the eastern Arctic. In H. Maschner,

O. Mason, & R. McGhee (Eds.), *The northern world AD 900–1400* (pp. 279–299). Salt Lake City: University of Utah Press.

Tape, K., Sturm, M., & Racine, C. (2006). The evidence for shrub expansion in Northern Alaska and the Pan-Arctic. *Global Change Biology,* 12: 686–702.

Terasmae, J., Webber, P. J., & Andrews, J. T. (1966). A study of the late-Quaternary plant-bearing beds in north-central Baffin Island, Canada. *Arctic,* 19(4): 296–318.

Tyrrell, J. B. (1898). The glaciation of north-central Canada. *Journ. Geology,* 6: 147–160.

Villarreal, S. (2013). *International Polar Year (IPY) Back to the Future (BTF): Changes in Arctic ecosystem structure over decadal time scales* (Doctoral dissertation). University of Texas at El Paso.

Villarreal, S., Webber, P. J., Johnson, D. R., Hollister, R. D., Lara, M. J., Lin, D. H., & Tweedie, C. E. (2014). Vegetation datasets from Northern Alaska, Baffin Island, Canada,and Beringia. In D. A. Walker (Ed.). Alaska Arctic Vegetation Archive (AVA) Workshop, Boulder, Colorado, Oct. 14–16, 2013. CAFF Proceedings Report 11. Akureyri, Iceland: Conservation of Arctic Flora and Fauna.

Walker, D. A. (Ed.) (2014). Alaska Arctic Vegetation Archive (AVA) Workshop, Boulder, Colorado, Oct. 14–16, 2013. CAFF Proceedings Report 11. Akureyri, Iceland: Conservation of Arctic Flora and Fauna.

Ward, W.H., & Baird, P.D. (1954). Studies in glacier physics on the Penny Ice Cap, 1953. Pt 1. *Journ. Glaciol..,* 2 (15): 342-355.

Watson, A. (2011). *A zoologist on Baffin Island, 1953: Four months of Arctic adventure.* Northants, U.K.: Paragon.

Webber, P. J. (1971). *Gradient analysis of the vegetation around the Lewis Valley, North Central Baffin Island, Northwest Territories, Canada* (Doctoral dissertation). Queen's University, Kingston, Ontario.

Webber, P. J. (2014). The nature and appropriateness to the Arctic Vegetation Archive (AVA) of data sets gathered using the Webber plant community sampling method. In D. A. Walker (Ed.), Alaska Arctic Vegetation Archive (AVA) Workshop, Boulder, Colorado, USA, Oct. 14–16, 2013. *CAFF Proceedings Report 11.* Akureyri, Iceland. ISBN: 978–9935–431–29–5.

Weber, H., Marmet, J., Röthlisberger, H., & Schwarzenbach, F. (2008). *Baffin Island: Arctic expedition, summer 1953* [Video]. Zug, Switzerland: Weissfilm.

Wilson, J. T., et al. (1958). Glacial map of Canada [Map]. Toronto: Geol. Assoc. Canada.

Wolken, G. J., England, J. H., & Dyke, A. S. (2005). Re-evaluating the relevance of vegetation trimlines in the Canadian Arctic as an indicator of Little Ice Age paleoenvironments. *Arctic,* 58(4): 342–353.

Wordie, J. M. (1938). An expedition to north west Greenland and the Canadian Arctic in 1937. *Geog. Journ.,* 92(4): 385–421.

MAJOR PUBLICATIONS INFLUENCED BY THE BAFFIN ISLAND EXPEDITIONS

Andrews, J. T. (1970). *A geomorphological study of post-glacial uplift with particular reference to Arctic Canada* (Inst. of British Geographers, Spec. Pub. No. 2: 1–156). London: The Alden Press, Oxford.

Andrews, J. T., (Ed.). (1985). *Quaternary Environments: Eastern Canadian Arctic, Baffin Bay and Western Greenland*. Boston, London and Sidney. Allen and Unwin.

Church, M. (1972). *Baffin Island sandurs: A study of Arctic fluvial environments* (Bull. No. 216). Ottawa: Geological Survey of Canada.

Ives, J. D., & Barry, R. G. (Eds.). (1974). *Arctic and alpine environments*. London: Methuen.

Fulton, J. R. (Ed.). (1989). *Quaternary Geology of Canada and Greenland*. Ottawa: Geological Survey of Canada.

LIST OF ABBREVIATIONS

ADM	Assistant Deputy Minister	IGU	International Geographical Union
AINA	Arctic Institute of North America	IGY	International Geophysical Year
asl	above sea level	IPY	International Polar Year
BP	Before Present	IHD	International Hydrological Decade
CAVM	Circumpolar Arctic Vegetation Map	LORAN	Long Range Navigation
CCGS	Canadian Coast Guard Ship	NORAD	North American Air Defense
DEW Line	Distant Early Warning Line	SAC	Strategic Air Command
DM	Deputy Minister	SWI	Summer Warming Index
DOT	Department of Transport	UBC	University of British Columbia
GSC	Geological Survey of Canada		

INDEX

NOTE: Page numbers in italics refer to photographs. Page numbers in bold refer to maps.

field facilities. *See* camp structures

Fielding, Tim, 78

Fife, Allan, 189

Fifth Thule Expedition, Danish (1921–1924), 9

firearms, 45

Fish, D. B., 78

Flint, Maurice, 9

Flint, Richard, 12–13, 52, 179–80, 182, 184

Flint Lake, 9, 35

Flitaway Lake, **10,** 11, 17, *31*, 48–49, *57*, 189, 208n6

Flitaway Lake base camp, **22,** 29–31, **49,** 51, **62,** 62–63, 70, **88**

Foley Island, 88

food, 38, 68, 208n12, 211n5 (chap. 4)

Fort Chimo, 50–51

fossils, marine, 90, 132, 179, 184–85

fossils, plant, 48, 189

Foxe, Luke, 9

Foxe Basin, 3, 7, 9, 27, *34*, *37*, 60, 89, 180, 184, 186, 188, 215n2

Foxe Dome, 184

Fraser, Keith, 29, 82, 142

Fremlin, Gerry, 74, 96

Freshney River, 27, 32, 208n5

Frobisher, Martin, 9

Frobisher Bay, 7, 12, 21, 51, 182, 207n4

Froese, Art, 125, 132

frost-shattered debris. *See* mountaintop detritus

fungi, 148

Fyles, J. G. (John), 13, 180, 213n6 (chap. 7)

G

Gamble, Sam, 45

Gander Aviation, 119

Gaudreau, Pierre, 61, 65–67

Gee Lake, 88

Generator Lake, 10, 125, *126*

Geodetic Survey Division, 125, 142

Geografiska Annaler, 210n7

Geographical Branch: location, 21; purpose and survival, 73, 75, 209n2; recruitment of women, 81; reorganization and demise, 2, 74, 138–39, 142; shield, 74, *75,* 151, 153, *153*. *See also* van Steenburgh, W. E.

Geographical Bulletin, 74, 82, 96

geography (scientific discipline), 1, 43, 75

Geological Survey of Canada. *See* GSC

geology (scientific discipline), 75

Geomorphology Section, 43

George River, 13

Gerasimov, Innokentiy, 76–77, 212n6

Gibbs Fiord, *134*

Gilbert, Robert, 85

glacial erosion, evidence of, 112, 178, 186, *198*

glacial features. *See* boulders; erratics; ice; ice-dammed lakes; marine terraces; moraines; pseudo-erratics; tors

Glacial Lake Agassiz, 3, 188

Glacial Lake Lewis, *14–15,* 24, **25,** *26,* 27, *27,* 183, 189

Glacial Lake McLean, 51

Glacial Lake Naskaupi, 51

glacial outwash plains. *See* sandar

glacier maps/mapping, 12–13, 86, 96, 182, 187

glaciology (scientific discipline), 43, 96, 137, 181

Glaciology Division. *See* Glaciology Section

Glaciology Section, 5, 43, 97, 181, 207n2 (Introduction), 209n6

Goldthwait, Richard, 12, 89

Goodfellow, Joan, 81

Goodison, Barry, 87, 95, 128, 160–61

Goudreau, Pierre, 62

Grant-Suttie Bay, 32, 89

Gray, Norman, 87, *175*

Greenland, 9, 178, 188, 191

Grenier, F., 213n6 (chap. 7)

GSC (Geological Survey of Canada), 12–13, 43, 46, 80, 212n10. *See also* Harrison, James M.

Gulf of Bothnia, 13, 179, 215nn2–3

H

Hainault, Robert, 87

Hall, Peter, 78

Hall Beach. *See under* DEW Line stations

Hamelin, Louis-Edmond, 44

Hamilton, Angus, 38

Hantzsch, Bernhard, 9, 182

Harrison, Dave (geographer), 67, 78, 87, 144, *171,* 189

Harrison, David (pilot): about, 91, 99; 1965 field season, 89–90, 92; 1966 field season, 99–102, 108, *109, 111,* 117, *118,* 119; 1967 field season, *127,* 132, 142, 144, 148, 151; helicopter flight "lessons," 92–93, 95

K

Karlén, Wibjörn, 85

Keewatin, 188

Keyhole Glacier, 119, *121–24*

Kihl, Rolf, 41, 51, 54, *55–56*, 61–63, 65–67

King, Cuchlaine: about, 168, *171*; 1965 fieldwork, 87–88, 95, 128; 1967 fieldwork, 145; other mentions, 61, 76, 81, 212n5

King Lake, *63*, 65, *66*

King River, 48, 52

King River base camp, 48, **49, 62,** 65, **88**

Koerner, Roy, 181, 183, 209n6

Krüger, Anne-Marie, 81

Kuujjuaq. *See* Fort Chimo

L

Labradorean Ice Sheet. *See* Laurentide Ice Sheet

Labrador Sea, 13, 51

Labrador-Ungava fieldwork, 13, 16, 20, 43, 80–81. *See also* McGill Lab; Torngat Mountains

Labrador-Ungava ice sheet, 13, 51, 188

Lamb, Greg, 36, 38–39

Lamothe, Claude: 1961 reconnaissance, 19, 21, 28–29, 35–36, *39*, 40; 1962 fieldwork, 54, *56*

Lancaster Sound, 9

Lara, Mark, 190

Laurentide Ice Sheet, 3, 9, 13, 51, 178–80, 186–88

Ledum groenlandicum (Labrador tea), 189

Levesque, Robbie, 29–35, 39–40

Lewis, Peter, 87, 125

Lewis, Vaughan, 24, 207n6

Lewis Glacier, 11, 31, *31*, 61, 189, 191

Lewis River, 11, 61–62, 65, 95, 189

Lewis River camp (1964), 78

Lewis River camp (1965), 87

"lichen-free" areas, 60, 64, 211n2 (chap. 5)

lichenometry, 4, 60, 79, 188–89, 211n1 (chap. 5). *See also* Beschel, Roland; Webber, P. J.

lichens and lichen cover, 24, 26, *26*, 27, 59–60, 64, 183, 188–89. *See also individual species*

Life magazine, 125, 132

light- and dark-toned areas, 13, 20, 24, 27, 182. *See also* lichens and lichen cover

Little Ice Age, 3, 183

Lloyd, T., 213n6 (chap. 7)

Logie, Jean, 119, *174*

Løken, Inger Marie, 81

Løken, Olav: about, 168–69, *171*; 1964 fieldwork, 76–79; 1965 fieldwork, 86–91; 1966 fieldwork, 119, 125, 135; 1967 fieldwork, 142, 144; other mentions, 5, 50, 96, 182, 184–86, 191

Longstaff, Tom, 133

Longstaff Bluff, 9, 11. *See also under* DEW Line stations

LORAN (long range navigation) sites, 11, 100

M

MacHattie, C., 89–90

Mackay, Ross, 44, 50, 96, 139, 213n6 (chap. 7)

Maktak Fiord, 10

Manning, Tom, 12, 182

maps and mapping. *See Atlas of Canada*; glacier maps/ mapping; submarine topography/mapping; Surveys and Mapping Branch; topographic maps/mapping

Marchant, P., 213n6 (chap. 7)

marine limits. *See* sea levels, former

marine terraces, 12, 179, 182, 215n4

Marmet, Jürg, 10

Marsden, Mike, 81

Mary River base camp, 61, **62,** 63

mass balance studies, 47–48, 88, 95, 125, 144–45

mass balance transect, Western Canada, 4, 85, 185

Mathiassen, Therkel, 182

Mattox, Bill, 50

Mattox, Joan, 50

The Maw, 117

McBeth Fiord, 119, 148

McCracken, Mary, 42–43

McEwen, Dave, 29, 39

McGill-Carnegie-Arctic Research Program, 207n1 (Introduction)

McGill Lab, 2, 20, 44, 208n3, 215n3. *See also* Labrador-Ungava fieldwork

McGill Sub-Arctic Research Laboratory. *See* McGill Lab

McLaren, Patrick, 145

mega-fossils. *See* fossils, marine

Meier, Mark, 86

Melville Peninsula, 182, 187

Mercer, John, 12, 182

Meteorological Branch, 45
meteorological stations, 95, 125
micro-fossils. *See* fossils, marine
Mid-Canada Line, 11
Miller, Gifford, 185, 188, 191, 215n10
Millest, Mary. *See* Strom, Mary
mollusc shells. *See* fossils, marine
moraines: cross-valley moraines, *14–15*, 27, *28*, 48, 60;
     end moraines, 12–13, 52, *54*, 101, *103–4, 106, 202*;
     ice-cored moraines, 43, 52, *54*, 62; lateral moraines,
     112, 132, 135, 186, *202*; medial moraines, *201–2*;
     moraine systems, 17, 20, 32, 187. *See also* Cockburn
     Moraines; Ekalugad Moraine
Moroz, G., 87
Morse, John, 78–79
mosses, 64, 148, 183, 189
mountaintop detritus, 110, 112, *113–14, 120*, 178–79, 186
Mount Asgard, 10
Mount Battle, 10
Mount Cook, 102
Mount Fleming, 10
Mount Longstaff, *133*
Müller, Fritz, 58, 77, 181
Murray, Lil, 73
Murray, Tom, 89, 101, 117, 142
Mya truncata, 90

N

names/naming of places and features, 5, 9–11, 81–82,
     207n4–207n5, 212n9, 212n11. *See also* Canadian
     Permanent Committee on Geographical Names
National Advisory Committee for Geological Research, 43
National Advisory Committee on Geographical Research,
     74, 96, 137–38, 140, 213n6 (chap. 7)
National Air Photo Library, 21, 85
National Atlas program, 74, 96
National Film Board (NFB), 212n3
National Museum of Natural Sciences, 189
Nicholson, Norman, 2, 21, 40, 44–45, 73–74, 182. *See
     also* Geographical Branch
nomenclature. *See* names/naming of places and features
North American Air Defense Command (NORAD), 11
North Atlantic Drift, 7

Northern Wings, 50, 56, 61
Norway, 7, 91, 185–86. *See also* Scandinavian hypotheses
Nuksuklorolu Mountain. *See* Inugsuin Pinnacles
Nuluujaak Mountain, 210n2
nunatak (term), 214n1 (chap. 11)
nunatak hypothesis, 112–13, 178, 185, 214n1 (chap. 11)

O

Odell, Noel, 178–79
Ommanney, C. S. L., 185
ookpiks, 148, 214n9
ordination analysis, 190
Orvig, Svenn, 58
O'Shaughnessy, Jim, 142
Østrem, Britta, 54, 209n5
Østrem, Gunnar: about, 169–70; recruitment to
     Geographic Branch, 43, 45; 1962 fieldwork, 50–52,
     54–55, *55–56*; 1963 fieldwork, 60–63; 1965
     fieldwork, 86–87, 95; other mentions, 4, 58, 77–78,
     85, 96, 125, 185, 209n5, 211n5 (chap. 4)

P

Pacific Coast Ranges. *See* Coast Ranges
Padloping Island, 135
Pan-American Institute of Geography and History, 74
Pangnirtung, 10
Papaver labradoricum (Iceland poppy), *154*
Parmelee, J. A., 148
Parry, W. E., 207n3
Pedicularis sudetica (Sudeten lousewort), *155*
Penny, William, 9
Penny Highlands, 186
Penny Ice Cap, 9–10, 12, 208n9 (chap. 1)
Pépin, Jean-Luc, 140–41
Perfection Pass, 101, *104*
permafrost, 181
Perraton, D. J., 78
Peterson, Jim, 78
Peyto Glacier, 4, 86, 185
Phillips, Bill, 148
Phillipson, Arthur, 214n7 (chap. 8)
Philpot, Jane, 87, 95, 119, 128, 132, 166–67, *173*